How Software Defined Networking (SDN) Is Going To Change Your World Forever

The Revolution In Network Design And How It Affects You

"An introduction to a new way of creating networks that will change everything"

Dr. Jim Anderson

Published by:
Blue Elephant Consulting
Tampa, Florida

Copyright © 2016 by Dr. Jim Anderson

All rights reserved. No part of this book may be reproduced of transmitted in any form or by any means, electronic or mechanical, including photocopying, recording or by any information storage and retrieval system without written permission of the publisher, except for inclusion of brief quotations in a review.

Printed in the United States of America

Library of Congress Control Number: 2016916499

ISBN-13: 978-1539310686
ISBN-10: 153931068X

Recent Books By The Author

Product Management

- What Product Managers Need To Know About World-Class Product Development: How Product Managers Can Create Successful Products

- How Product Managers Can Learn To Understand Their Customers: Techniques For Product Managers To Better Understand What Their Customers Really Want

Public Speaking

- Tools Speakers Need In Order To Give The Perfect Speech: What tools to use to create your next speech so that your message will be remembered forever!

- How To Create A Speech That Will Be Remembered

CIO Skills

- Becoming A Powerful And Effective Leader: Tips And Techniques That IT Managers Can Use In Order To Develop Leadership Skills

- CIO Secrets For Growing Innovation: Tips And Techniques For CIOs To Use In Order To Make Innovation Happen In Their IT Department

IT Manager Skills

- Save Yourself, Save Your Job – How To Manage Your IT Career: Secrets That IT Managers Can Use In Order To Have A Successful Career

- Growing Your CIO Career: How CIOs Can Work With The Entire Company In Order To Be Successful

Negotiating

- Learn How To Signal In Your Next Negotiation: How To Develop The Skill Of Effective Signaling In A Negotiation In Order To Get The Best Possible Outcome

- Learn The Skill Of Exploring In A Negotiation: How To Develop The Skill Of Exploring What Is Possible In A Negotiation In Order To Reach The Best Possible Deal

Note: See a complete list of books by Dr. Jim Anderson at the back of this book.

Acknowledgements

Any book like this one is the result of years of real-world work experience. In my over 25 years of working for 7 different firms, I have met countless fantastic people and I've been mentored by some truly exceptional ones. Although I've probably forgotten some of the people who made me the person that I am today, here is my attempt to finally give them the recognition that they so truly deserve:

- Thomas P. Anderson
- Art Puett
- Bobbi Marshall
- Bob Boggs

Dr. Jim Anderson

This book is dedicated to my wife Lori. None of this would have been possible without her love and support.

Thanks for the best years of my life (so far)...!

Speaking.　Negotiating.　Managing.　Marketing.

Table Of Contents

1 INTRODUCTION: THE WAY THAT COMPUTER NETWORKING IS DONE TODAY 12
    1.1 THE PROBLEM WITH THE WAY THAT NETWORKS ARE DONE TODAY .12
    1.2 A SHORT HISTORY OF SDN 14
    1.3 HOW ARE TODAY'S NETWORK ROUTERS DESIGNED? 15
    1.4 HOW IS COMPUTER NETWORKING DONE TODAY? 16
    1.5 THE CORRECT WAY TO CREATE A NETWORK CONTROL PLANE 20
        1.5.1 The Forwarding Function 22
        1.5.2 Network State Function 23
        1.5.3 The Configuration Function 28
    1.6 WHAT IS SDN'S "KILLER APP"? 34

2 THE IMPORTANCE OF SDN 36

3 IMPACT OF SDN ON THE DESIGN OF FUTURE NETWORKS 37
    3.1 SEPARATION OF THE CONTROL AND DATA PLANES 37
    3.2 THE ARRIVAL OF EDGE-ORIENTATED NETWORKING 39
    3.3 IN CONCLUSION 42

4 AN EXAMPLE OF HOW SDN WOULD WORK IN A NETWORK 43

5 HOW TELECOM SERVICE PROVIDERS VIEW SDN 46
    5.1 ATTRIBUTES OF SUCCESSFUL TELECOM SDN NETWORKS 48
    5.2 HOW SERVICE PROVIDERS PLAN ON INTRODUCING SDN INTO THEIR NETWORKS 51
    5.3 SUMMARY 60

6 THE DEVELOPER AND THE NETWORK 61
    6.1 THE PROBLEM WITH TODAY'S APPLICATIONS AND NETWORKS 62
    6.2 WHAT IS A PROGRAMMABLE NETWORK? 64
    6.3 THE VALUE OF HAVING MORE NETWORK / APPLICATION INFORMATION 65

	6.4	EVOLVING LEGACY NETWORKS TO SDN NETWORKS 66
7		**EXAMPLES OF NETWORKED APPLICATIONS THAT CAN ONLY BE OFFERED IN AN SDN NETWORK 70**
	7.1	SERVICE ENGINEERED PATHS 71
	7.2	SERVICE APPLIANCE POOLING 71
	7.3	CONTENT REQUEST ROUTING 71
	7.4	BANDWIDTH CALENDARING 72
	7.5	SOCIAL NETWORKING 73
8		**HOW VENDORS ARE RESPONDING TO SDN 75**
	8.1	CISCO'S RESPONSE TO SDN 75
	8.2	VMWARE'S NSX SDN 80
	8.3	JUNIPER 82
	8.4	OPENDAYLIGHT SDN 85
	8.5	BIG SWITCH NETWORKS 88
9		**GOOGLE AND SDN 89**
	9.1	GOOGLE'S PROBLEM WITH ITS NETWORK 90
	9.2	GOOGLE'S MOTIVATION TO FIND A BETTER NETWORKING SOLUTION 93
	9.3	THE IMPORTANCE OF NETWORK TESTING 100
	9.4	SIMULATING THE GOOGLE WAN 102
	9.5	WHY GOOGLE IS INTERESTED IN SDN... 103
	9.6	GOOGLE'S OPEN FLOW WAN 105
		9.6.1 *Google's G-Scale Network Hardware 107*
		9.6.2 *How Google Rolled Out Their SDN Network 108*
		9.6.3 *Bandwidth Brokering and Traffic Engineering In Google's Backbone Network 116*
		9.6.4 *Results Of Google's Deployment Of SDN 122*
		9.6.5 *Challenges Of Implementing A SDN Network 127*
		9.6.6 *Google's Conclusions After Having Implemented A SDN Network 129*
10		**OPENFLOW TOPICS 131**
	10.1	AN OVERVIEW OF THE OPENFLOW SWITCH SPECIFICATION 133
		10.1.1 *OpenFlow Ports 136*

    10.1.2   OpenFlow Flow Tables 136
    10.1.3   The OpenFlow Channel 146
  10.2   OPENFLOW CONFIGURATION AND MANAGEMENT PROTOCOL......... 155
    10.2.1   Setting Up A Connection Between A Switch And A Controller 157
  10.3   THE CONFORMANCE TEST SPECIFICATION FOR OPENFLOW SWITCH SPECIFICATION 1.0.1... 159
  10.4   THE OPENFLOW™ CONFORMANCE TESTING PROGRAM 162

11   THE FUTURE OF SDN 164

  11.1   HOW CAN SDN BE ADDED TO EXISTING ENTERPRISE NETWORKS? . 165
  11.2   CAN SDN BE USED WITH TRANSPORT NETWORKS?...................... 169
  11.3   HOW CAN SDN CONCEPTS BE EXTENDED TO WORK WITH OPTICAL TRANSPORT NETWORKS? 172
  11.4   CAN A WAN NETWORK'S UTILIZATION BE INCREASED BY USING SDN TECHNOLOGY? 176
  11.5   HOW SCALABLE ARE SOFTWARE DEFINED NETWORKS? 184
    11.5.1   SDN Scalability Issues 184
    11.5.2   Controller Designs For Scalability 185
    11.5.3   Potential SDN Network Scalability Issues 187
    11.5.4   Network Types 191
    11.5.5   Next Steps 192
  11.6   NETWORK MANAGEMENT LANGUAGES FOR SOFTWARE DEFINED NETWORKS... 193
    11.6.1   How To Query The Network State Using Frenetic 195
    11.6.2   How To Create Network Policies Using Frenetic 197
    11.6.3   How To Perform Consistent Updates Using Frenetic 198
  11.7   CREATING A SDN CONTROLLER THAT IS BOTH ELASTIC AND DISTRIBUTED 199
    11.7.1   How To Migrate Switches Using The ElastiCon Distributed Controller 203
    11.7.2   Load Adaptation 207

12   CONCLUSION 210

13   REFERENCE... 214

List Of Figures

FIGURE 1: ARCHITECTURE OF A MODERN NETWORK ROUTER 16
FIGURE 2: THE 3 PLANES OF A TRADITIONAL ROUTER 17
FIGURE 3: DATA PLANE ABSTRACTION LAYERS 18
FIGURE 4: ROUTER COMPONENTS IN AN SDN NETWORK 21
FIGURE 5: NETWORK OF SWITCHES 24
FIGURE 6: TRADITIONAL CONTROL MECHANISMS 25
FIGURE 7: SOFTWARE DEFINED NETWORK (SDN) 26
FIGURE 8: SDN WITH GLOBAL NETWORK VIEW & CONTROL PROGRAM 27
FIGURE 9: A SDN EXAMPLE - ACCESS CONTROL 30
FIGURE 10: SDN EXAMPLE - ACCESS CONTROL ABSTRACT VIEW 30
FIGURE 11: COMPONENTS OF A SDN 31
FIGURE 12: SAMPLE SDN NETWORK 43
FIGURE 13: SDN CONTROLLER UPDATING ROUTERS 44
FIGURE 14: APPLICATION MOVEMENT IN AN SDN NETWORK 45
FIGURE 15: TELECOM SERVICE PROVIDER'S SDN MOTIVATION 46
FIGURE 16: DEPLOYMENT SCENARIOS FOR CARRIER SDN 50
FIGURE 17: HOW TRAFFIC STEERING IS DONE TODAY 56
FIGURE 18: USING OPENFLOW FOR TRAFFIC STEERING OPTIMIZATION 57
FIGURE 19: EXAMPLE OF BANDWIDTH-ON-DEMAND FOR A HYBRID CLOUD
    ARCHITECTURE 59
FIGURE 20: HOW APPLICATIONS AND THE NETWORK WILL INTERACT 67
FIGURE 21: PROPOSAL FOR ADDING SDN WITHOUT BREAKING THE INTERNET 68
FIGURE 22: ADDING SDN TO A LEGACY ROUTER 69
FIGURE 23: CISCO ACI FABRIC 77
FIGURE 24: VMWARE'S NSX VIRTUALIZATION PLATFORM 81
FIGURE 25: JUNIPER'S CONTRAIL CONTROLLER 84
FIGURE 26: TECHNICAL OVERVIEW OF THE OPENDAYLIGHT CONTROLLER HYDROGEN
    RELEASE [SOURCE: OPENDAYLIGHT PROJECT] 86
FIGURE 27: THE BIG SWITCH NETWORKS OPEN SDN 89
FIGURE 28: SAMPLE NETWORK WITH 3 APPLICATION TRAFFIC FLOWS 93
FIGURE 29: NETWORK WITH LINK FAILURE 94
FIGURE 30: NETWORK WITH ONE TRAFFIC FLOW REPAIRED 95

FIGURE 31: NETWORK AFTER ALL 3 TRAFFIC FLOWS HAVE BEEN REBUILT 95
FIGURE 32: USING CENTRALIZED TRAFFIC ENGINEERING TO REBUILD THE NETWORK 97
FIGURE 33: REBUILT NETWORK WITH CENTRALIZED TRAFFIC ENGINEERING............98
FIGURE 34: GOOGLE DATA CENTER LOCATIONS 106
FIGURE 35: GOOGLE'S G-SCALE BACKBONE NETWORK WAN DEPLOYMENT 108
FIGURE 36: MIXED SDN DEPLOYMENT 109
FIGURE 37: MIXED SDN DEPLOYMENT WITH CONTROL APPLICATIONS 111
FIGURE 38: ADDING PROTOCOLS TO ALLOW SERVER TO COMMUNICATE WITH NETWORK...113
FIGURE 39: ADDING ADDITIONAL FUNCTIONALITY TO SDN NETWORK 114
FIGURE 40: NETWORK BANDWIDTH REQUESTS FOR COMMUNICATION BETWEEN APPLICATIONS (IN MBPS)...116
FIGURE 41: GOOGLE'S SDN NETWORK HIGH-LEVEL ARCHITECTURE 118
FIGURE 42: ARCHITECTURE OF THE BANDWIDTH BROKER 119
FIGURE 43: TRAFFIC ENGINEERING SERVER'S SERVICE ARCHITECTURE 120
FIGURE 44: ARCHITECTURE OF THE SDN CONTROLLER 121
FIGURE 45: USE OF THE SDN CONTROLLER IN GOOGLE'S BACKBONE NETWORK 122
FIGURE 46: WAN CONVERGENCE UNDER FAILURE CONDITIONS......................126
FIGURE 47: OPENFLOW SWITCH MAIN COMPONENTS [20] 135
FIGURE 48: OPENFLOW PACKET FLOW THROUGH THE PACKET PROCESSING PIPELINE [20] 138
FIGURE 49: OPENFLOW PACKET MATCHING PROCESS......................................139
FIGURE 50: OPENFLOW CONTROLLER MODES...151
FIGURE 51: AUXILIARY CONNECTIONS TO AN OPENFLOW SWITCH 154
FIGURE 52: RELATIONSHIP BETWEEN OF-CONFIG PROTOCOL AND THE OPENFLOW PROTOCOL 156
FIGURE 53: TEST BED USED TO EXECUTE OPENFLOW TEST CASES 161
FIGURE 54: 3 POSSIBLE ENTERPRISE NETWORK SDN DEPLOYMENT SCENARIOS ..166
FIGURE 55: ARCHITECTURE OF SWAN [33] 181
FIGURE 56: SDN NETWORK FLOW SETUP PROCESS..188
FIGURE 57: HOW A SDN NETWORK CONVERGES ON A LINK FAILURE...............190
FIGURE 58: ARCHITECTURE OF THE ELASTICON DISTRIBUTED CONTROLLER 201

1 Introduction: The Way That Computer Networking Is Done Today

The legacy networks that are in use by organizations today have an infrastructure that is typically a mix of equipment from multiple vendors, platforms and different protocol solutions. This makes the ultimate goal of creating an integrated network ecosystem a difficult if not impossible process for many organizations.

The arrival of Software Defined Networking (SDN) is an approach to building networks using open protocols, such as OpenFlow, that allow globally aware software control to be applied at the edges of the network in order to access network switches and routers that typically would use closed and proprietary firmware.

Software Defined Networking is not a revolutionary new technology. Instead, it is better to think of this as being a new way of organizing computer network functionality. SDN allows the network to be virtualized. That's where the real power of SDN comes from.

1.1 The Problem With The Way That Networks Are Done Today

Today's networks are a result of how the future was envisioned to be when a number of key protocol decisions were made back in the 1970's. It was envisioned that 32 bits would be big enough to handle all of the IP addresses that the Internet would ever need (roughly 16 million) – and it was until February of 2011 when the IP4 addresses ran out [21]. Networks were also envisioned to be fixed things: once set up, they would not change very much. This situation has changed greatly.

The servers that connect to today's networks have undergone a dramatic transformation in the past few years. The arrival of server virtualization has fundamentally changed what it means to be a server. Almost overnight, severs have become very dynamic in terms of how they are created and where they are located. Additionally, the number of servers that want to use the network has also increased.

Before the arrival of large scale virtualization of servers, applications used to be associated with a single server that had a fixed location on the network. These applications then spent their time using the network to exchange information with other applications that were similarly in fixed locations.

Today, everything has changed. Applications can be distributed across multiple virtual machines (VMs). Each of these VMs can then exchange traffic flows with each other. Network managers can move VMs in order to optimize and rebalance server workloads. This movement of applications can then cause the physical end points of an existing flow to change. The ability to migrate VMs creates challenges to many aspects of so-called traditional networking including addressing schemes and namespaces along with the basic notion of a segmented, routing-based design [19].

As though just the virtualization of servers wasn't enough to make things complicated in a network, now many companies are also using a single network to deliver all of the voice, video, and data networking needs that they have. In today's legacy networks, there is a concept of QoS that can be used to provide a differentiated level of service for different applications.

However, the provisioning of those QoS tools is highly manual. Network staff have to configure each vendor's equipment separately, and adjust parameters such as network bandwidth and QoS on a per-session, per-application basis [19]. The way that legacy networks are created is basically static and this means that the network cannot dynamically adapt to changing traffic, applications, and user demands

1.2 A Short History Of SDN

All of the work that has led to what we now know as SDN can be traced back to work that was performed in 2004. This research was a search for a new network management paradigm. It took on two different forms. The Routing Control Platform (RCP 40) work that was done at Princeton and Carnegie Mellon University. At the same time the network security work was being done as a part of the SANE Ethene project at Stanford University and the University of California at Berkeley .

This work was then built on in 2008 by two different groups. The startup Nicira which was eventually bought by VMWare created the Network Operating System (NOX). At the same time, Nicira worked with teams at Stanford University to create the OpenFlow switch interface.

By the time that 2011 arrived, the defacto standards body of the SDN space, the Open Networking Foundation, already had broad networking industry support. It had 69 members who consisted of firms such as Cisco, juniper, HP, Dell, Broadcom, IBM, etc. Its board consisted of members who were drawn from some of the networking industry's biggest firms: Google, Verizon, Yahoo, Microsoft, DeutchTelecom, Facebook, and NTT.

One of the most significant events in the history of the SDN technology occurred in 2012. At the Open Networking Summit which was attended by over 1,000 engineers from the networking industry, Google made the announcement that they were already using SDN technology. Google said that they were using SDN as a part of the Wide Area Network (WAN) that is used to interconnect their data centers [5].

1.3 How Are Today's Network Routers Designed?

Today's networks are built by interconnecting 10's, 100's, and sometimes 1,000's of sophisticated network routers whose sole purpose is to accept data packets from applications and to then forward them to the next router which is just a little bit closer to the data packet's eventual end destination.

Figure 1 shows a high level view of the components of a modern network router. A router is provided by a vendor who has gone to great lengths to create a unified stack of functions that all work together. This functionality starts with the specialized packet forwarding hardware whose job it is to accept data packets and then forward them on to the next router in order to eventually deliver them to the destination application. Controlling all of this is the router's operating system. The operating system is the router vendor's "crown jewels" and has been optimized to work with the vendor's underlying hardware platform. On top of all of this are the various network applications, such as specialized routing protocols which use the operating system and the packet forwarding hardware to accomplish their goals.

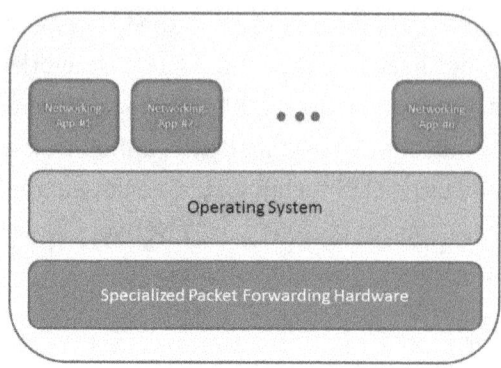

Figure 1: Architecture Of A Modern Network Router

In order to make changes to how the network behaves, it is necessary to access each router in the network and then by issuing a set of operating system commands in the language that has been defined by the router vendor, how the router behaves can be changed. This is a closed environment and does not easily allow the router to easily interact with any other components that make up the network.

1.4 How Is Computer Networking Done Today?

There would be no reason to change the way that we do computer networking to use SDN or any other new organization of network functionality if we weren't having some problems with the way that we do computer networking today. It would probably be a good idea if we started our discussion by first taking a look at how computer networking is being done today.

Computer networking today uses three separate planes to accomplish tasks: the data plane, the control plane, and the management plane as shown in Figure 2. The data plane is tasked with the job of processing packets that are received. When a packet arrives, the data plane functionality uses its information about the local forwarding state and the information that is contained in each packet's header to make a decision about whether to drop the packet or forward it. If you decide to forward it, then you have to decide which computer and which port the packet should be sent on to. In order to keep up with all of the packets that are arriving, the data plane processing must be done very, very quickly. The management plane is responsible for coordinating the interaction between the control plane and the data plane.

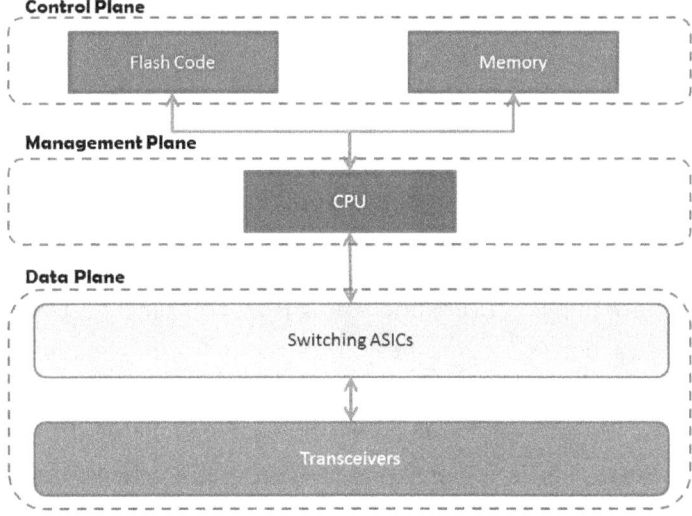

Figure 2: The 3 Planes Of A Traditional Router

The control plane is what computes the forwarding state that the data plane uses to forward the packet. This forwarding state can be calculated in a number of different ways: distributed algorithms, centralized algorithms, or it can be manually configured.

The data plane and the control plane are very different. They have both been designed to accomplish different jobs.

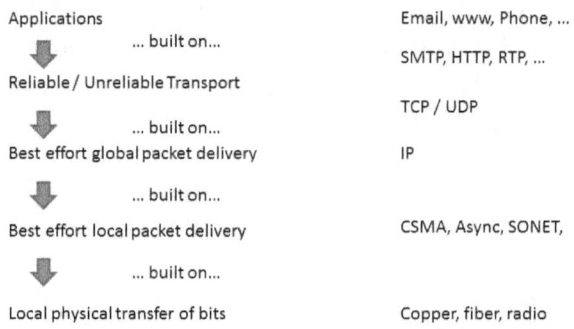

Figure 3: Data Plane Abstraction Layers

Figure 3 shows the way that the data plane has been constructed using abstractions. An unreliable set of transport layers have been used to ultimately create a reliable transport layer.

Where things get very interesting is that no similar set of abstractions exists for the control plane. There is no underlying simplification that has been used to build a control plane. Instead, what has been done is that a large number of mechanisms have been created in order to implement different types of control planes.

The control plane mechanisms that have been created have all been designed in order to accomplish a variety of different goals. One example of this is the routing control planes that implement a wide variety of distributed routing algorithms. There are also isolation control planes that can be used to provide Access Control Lists (ACL), virtual local area networks

(VLAN), Firewalls, etc. Finally, there are traffic engineering based control planes that can use adjusting link weights to make routing decisions, implement multi-protocol label switching (MPLS), etc. All of these different approaches are trying to affect the routing of packets – they want to control how the forwarding state is calculated.

The problem with the control planes that are being used in today's networks is that they are not modular – you can't use them together. Each one of them solves part of the problem of controlling a network, but none of them solve all of the problems of controlling a network. Effectively each one of them provides limited functionality.

The modern network control plane provides far too many mechanisms but with precious little functionality. The reason for this is because there has been no way to abstract just exactly what the control plane does. Every time we run into a new problem (e.g. people are illegally gaining access to our network) you end up defining a whole new solution from scratch. All of these new solutions end up only implementing some of the functionality that is required – no single solution has all of the required control plane functionality.

The end result of designing the control plane this way is that there is a disharmony created between the enterprise network and the enterprise's business needs. A simple example of this would be in data route selection in today's legacy networks. If an application has a great deal of data that needs to be transferred to another application, it may end up using the fastest and most costly link in the network simply because it is available. However, if the application was able to "talk" to the network, it might discover that there was a slower and less expensive link that would serve its purposes just as well. However, the design of the control plane that is being used in today's networks does not allow this type of information to be shared.

1.5 The Correct Way To Create A Network Control Plane

The control plane is designed to compute the forwarding state under three different constraints. These 3 constraints are:

1. It has to be consistent with the low level hardware and software. You need to know what the ASIC is – you need to know what the hardware and the software are doing in the switch.

2. You have to base it on the entire network topology.

3. The forwarding state has to be inserted into every single physical forwarding box in the network.

Every time a new network protocol is designed we come back to these three problems and create yet another solution to them. It turns out that this is a very bad way to go about doing things.

> **Programming Analogy**: Pretend that you had been told that you had to write a program. Your program has the following requirements (1) it has to be aware of the hardware that it is running on (registers, limited available operations, etc.) and (2) you had to specify where each bit would be stored.

A programmer would not be willing to deal with this type of low-level complexity for long. Very quickly abstractions of the hardware system would be created. First, a programming language that was independent of the physical hardware would be created (compiler) and then a virtual memory interface (operating system) would be created.

Programmers have learned to use abstractions to separate the solutions that they create from the real-world concerns of the systems that their solutions are being implemented on. This is exactly what network designers should do also!

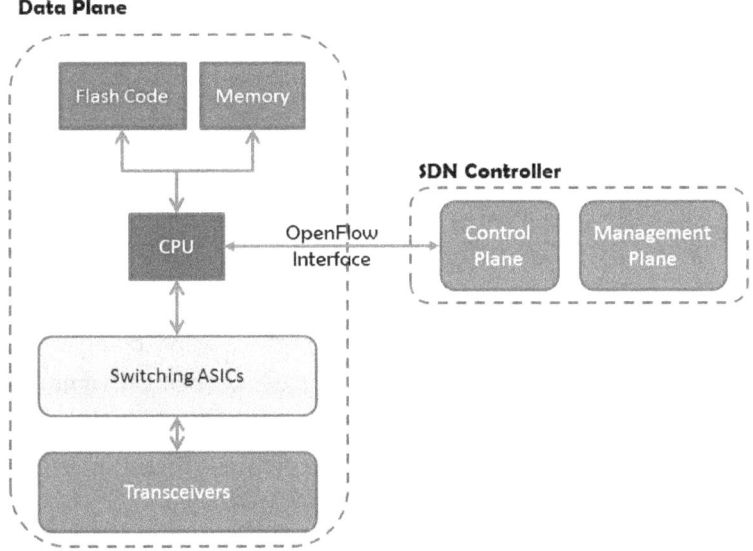

Figure 4: Router Components In An SDN Network

Figure 4 shows the different architecture that a router in an SDN network would have. The data plane functionality would be all that the router hardware would provide. The control plane and the management plane would be provided by a software application that was executing on a separate platform which was connected to the router via a secure data link.

In order to define functions for the network's control plane, what functions are needed? The first would be a type of general forwarding model that could hide the details of the low-level hardware / software that has been used to create the network switch. Next, a function to determine the current network state would be needed. This would be used to allow decisions to be made based on the entire complicated network. Lastly, a way is needed to actually configure the network. The goal is to avoid having to configure every physical box in a network (there could be thousands or even millions of boxes in a single network) therefore you need a function that allows you to simplify configuration. This can be done by computing the configuration of every physical device.

1.5.1 The Forwarding Function

When thinking about the best way to design a network control plane, one of the first functions that we want to implement is the packet forwarding function. If we were able to ignore how this functionality has been done in the network switches that are deployed in today's existing networks, it becomes clear that the forwarding function needs to be implemented independently of how the network switch is actually implemented. Our goal is to be able to express what we want to have to happen to a packet without having to worry about what switch is going to be used to implement it.

What this means is that the network switch should be able to use any set of ASICs with varying degrees of capabilities and this should have no effect on the forwarding function. Additionally, the software that is executing on the network switch could be from any vendor and this too would have no impact on the forwarding function.

In the constantly evolving world of Software Defined Networking, the OpenFlow interface is one proposal for how centralized control plane software would communicate with a network switch. Note: Vendors such as Cisco and VMWare have come up with their own alternative proposals. The OpenFlow interface is a set of application programming interfaces (APIs) that would allow an external software application to communicate with a network routing switch. The network packet that they would be discussing would be called a flow entry. A flow entry takes the form of a packet template and an action and is defined like this: <header, action>. A flow entry says that if a packet matches the template, then the switch should take the action (e.g. drop, forward out a specific port, etc.). OpenFlow is simply a general language that every switch in the network has to understand.

At a high level, the idea of an OpenFlow interface is very easy. However, when the details start to get examined, things become considerably more complex. In order to implement OpenFlow, designers need to make a lot of different decisions. These can include how best to perform rapid header matching. They can also include what actions are going to be allowed to be taken once a header has been matched.

1.5.2 Network State Function

In order to create a network state, the first thing that you want to do is to find a way to "abstract away" all of the complicated distributed functionality that is going to be required in order to collect the information that is necessary to create the network state.

The ultimate goal of the network state function is to present a "global network view". What this would look like would be a graph (objects and links) that has network information associated with it: network delay, link capacity, recent loss rate. Once the network graph was created, then the controlling software would be able to make decisions about what it wanted to do based on the graph. If access to the graph was provided via an API, then the actual network elements that make up the network could then be controlled via the API.

The global network view functionality could be implemented as part of a network operating system. This software would run on servers that were separate from the switches that were used in the network. It could, of course, be replicated in order to increase its reliability.

In order to keep all of the network data current, information would have to flow bidirectionally between the network operating system software and the network servers. This information flow would allow the global network view to be constantly updated with a view of what was happening at each switch in the network. Likewise, each switch in the network could then be updated in order to provide accurate control of packet forwarding.

1.5.2.1 Traditional Network Design

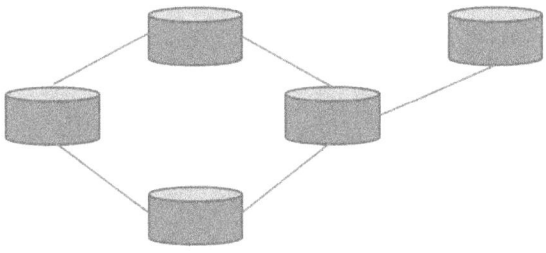

Figure 5: Network of Switches

Figure 5 shows a typical network. This network consists of a set of network switches that have the job of routing packets of user information between them. Connections exist between some of the switches in the network, but not all. However, every switch is reachable via one or more paths.

These connections are also used to allow each of the network switches to exchange control information which they can then use to update their individual "view" of the network. This in turn is then used to modify how they choose to forward the packets that they receive.

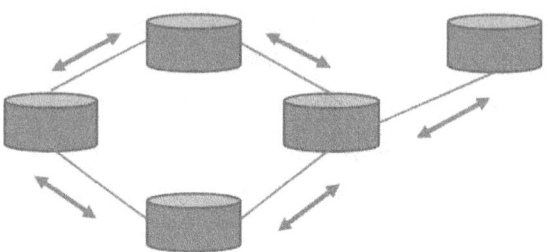

Figure 6: Traditional Control Mechanisms

Figure 6 shows an updated version of the network that was shown in Figure 5. Now more detail has been added on the control mechanisms that are being used in this network. Standards are being used to implement a peer-based routing algorithm. This algorithm is, by necessity, is a distributed algorithm that runs between neighbor switches in the network.

Because of the way that networks are designed and implemented today, this control algorithm is both complicated and task-specific. As an example, the control mechanism could be a shortest path first (SPF) algorithm that has been customized to allow the calculation of two disjoint paths which will be used for packet forwarding.

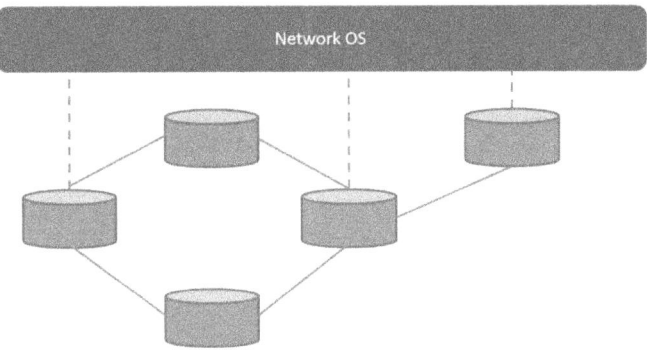

Figure 7: Software Defined Network (SDN)

Figure 7 shows how today's complicated network design can be simplified by the new approach that SDN offers. In this network, there is a general purpose software algorithm that is now running on the Network OS servers. This software talks to each of the switches in the network in order to determine the topology of the network. Once this information has been collected, the Network OS software is then able to create a global network view, as shown in Figure 8. The key characteristics of this piece of software is that it is both extensiable and flexible.

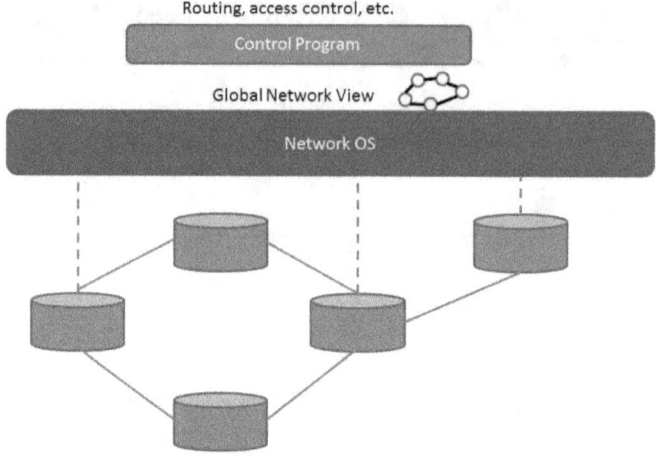

Figure 8: SDN With Global Network View & Control Program

Once a global network view is available, it is then possible to create a control program that operates at a higher layer. The control program can take on many different forms: it could be a routing program, an access control program, traffic engineering, etc. All of these different types of control programs can use the global network view to make packet forwarding decisions for the network. Note that this centralized decision making based on a global state is very different from today's networks in which distributed decisions are made based on imperfect local knowledge of the network's global state.

1.5.2.2 SDN Requires A Major Change In How We View Network Control

When SDN is being used to control a network, how the control program operates is much different from how routing is managed in a traditional network. The way that the information that will be used to determine how each of the switches in the network should forward packets is now going to be determined based on the global network view.

The way that the network determines how to forward packets that are traveling across it has been radically changed. Instead of using a distributed algorithm in which each switch in the network did its own set of calculations, now there is a centralized control program that has that responsibility.

This control logic runs as a program on the network OS and uses an API to interact with the global network view. Ultimately, the best way to envision the control program is to view it as being a type of graph algorithm.

This changes how things will be done when there is a need to do something different with the network in terms of how packets are routed. Instead of having to once again redesign a distributed routing algorithm, instead only the centralized control program will have to be modified. This makes the way that packets are forwarded much easier to verify, maintain, and to investigate.

1.5.3 The Configuration Function

The control program is where the behavior that is desired for packets that are being routed is going to be expressed. The underlying network by itself will have no idea how packets are to be routed. As an example, if it is determined that packets coming from node A should never travel though node B, then the control program will be responsible for expressing this. The underlying network would never have a way of telling if packets from A should or should not be routed through node B.

The control program will not be responsible for implementing the routing behavior on a physical network infrastructure. In the example, the control program will not be responsible for implementing the rule that node A should not talk to node B.

Instead, that will require detailed configuration information be placed in the forwarding tables in every router along every path in the network.

In order to implement this separation between the control program and the implementation of the output of the control program, in SDN the control program only interacts with an abstraction of the underlying network in which many of the details have been removed.

The global network view graph that the control program would interact with will only have enough detail so that the control program will be able to express its goals ("Node A's packets will never travel through Node B"). The amount of detail that the global network view will provide to the control program will depend on the function that the control program is trying to do. Quality of service, access control, or even traffic engineering would all require different amounts of detail on the underlying network.

A good way of thinking about this concept is to relate it to the world of programming languages and compliers. The computer that will eventually be used to run the program has a specific instruction set and eventually the program that the programmer writes will have to be expressed in that language. However, when the programmer is writing the program, they don't have to worry about the level of detail that goes along with writing a program in the computer's native language. Instead, they can use the higher-level language and not have to worry about all of the low-level details of how to get the program to run on the actual hardware –the compiler will take care of this.

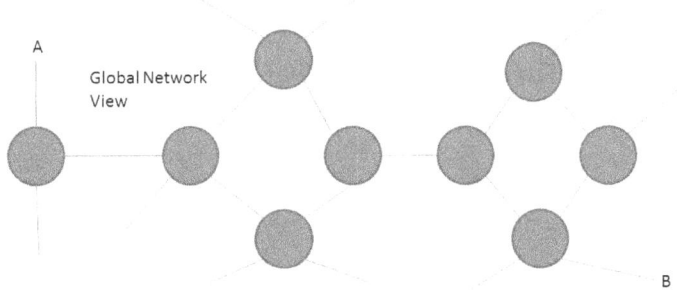

Figure 9: A SDN Example - Access Control

As example of this, in Figure 9, a sample network is shown. Assuming that a control program is executing an access control function, it will want to prevent Node A from communicating with Node B. In a traditional network, the control program would have to now perform the computations to determine how routing would be permitted to be done in this network and then it would have to go and configure the routing tables on every node in the network in order to implement the results of that routing algorithm.

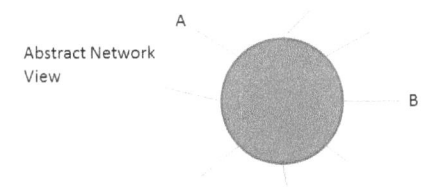

Figure 10: SDN Example - Access Control Abstract View

An alternative way of implementing the access control function is to present the algorithm with the view of the network shown in Figure 10. In this figure, only the external interfaces on the network are shown to the algorithm.

The access control algorithm is then asked the question "which nodes should be allowed to talk to which nodes?" Once the access control functions identifies that Node A should not be able to route packets through Node B, then its task is complete.

It will now be the responsibility of the "compiler" to make sure that all of the correct packet routing information is placed into the correct packet forwarding tables in each of the switches that make up the network.

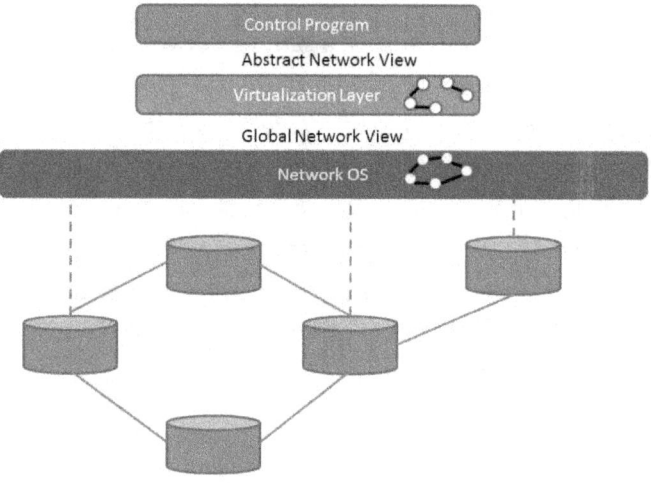

Figure 11: Components Of A SDN

In order to allow the "compiling" of the control program's packet routing decisions to be translated into actual configuration commands for each of the switches in the physical network, a new layer has to be added to the Software Defined Networking model. This layer is called the Virtualization Layer.

One of the jobs of the Virtualization Layer is to present an abstract network view to the control program – a view that has all of the unnecessary details removed. The control program will "see" a very simple view of the network. In our example, the control program will then decide that it wants traffic from Node A to not be routed through Node B.

The Virtualization Layer will then take the access control decisions made by the control program and converts it to the global network view. The updated global network view is then handed off to the network OS which then has the responsibly for properly configuring each of the switches that make up the actual physical network.

One additional point to make here is that the underlying hardware no longer has to be a sophisticated router. Instead, a relatively simple device that can provide basic packet forwarding hardware will do. This simple device will communicate with the Network OS using a protocol such as OpenFlow.

1.5.3.1 Separation Of Functionality

The architecture of a SDN based network has a very clean separation of functionality. The control program portion of the network has one very straightforward job to do: express the operator's requirements for how packets are to be routed in the network. The functionality of the control program is driven by operator requirements. When operator requirements change, the control program is the only portion of the SDN network that has to be changed.

The Virtualization Layer is built to take the abstract view that is presented to it by the control program and translate it into the more detailed global view. It is able to do this because its actions are driven by the specification abstraction for the particular task that it is doing.

Finally, the Network Operating System takes the packet routing control commands that have been created based on the global network view and maps them to the physical switches that make up the network that is being controlled. The API that the Network Operating System uses to talk to each of the network switches is driven by the network state abstraction. The interface between the Network Operating System and the switches is driven by the forwarding abstraction – potentially the OpenFlow interface.

Note that if there are changes in the underlying physical network, then these changes will travel up to the Network Operating System. Once there, they'll be reflected in the Global Network View and eventually in the Abstract Network View.

Working together, the Network Operating System, the Virtualization Layer, and the Control program make up the layers for the SDN control plane. Just as the network data plane has always had layers to implement its functionality, in an SDN based network, now the control plane has layers to implement its functionality.

1.5.3.2 Network Complexity Still Exists

The use of a SDN architecture does not make creating a properly operating network any easier than it is today. However, what it does do is to take the complexity associated with this task and move it around. The hard parts are placed where they should be.

The Network Operating System and the Virtualization layer components of a SDN network remain complex pieces of software. The functionality that these layers have to accomplish requires them to both be large pieces of software and to have a great deal of logical complexity designed into them.

However, a SDN designed network is able to accomplish two main objectives. The first is that it is able to simplify the interface for the control program portion of the control plane. This portion of the network can now be made user specific. The hard part of creating a properly operating network, the complex code, can now be pushed into the reusable SDN platform piece of the design.

Note that this design is very similar to how compliers for software languages are designed. Compliers are very hard to create. They contain a great deal of complexity and they have to have a very good understanding of the physical computer that the program will eventually be running on. However, once a correctly working compiler has been created, then the hard work has been done and network designers can just worry about stating what they want to have happen and not have to worry about how it will happen.

1.6 What Is SDN's "Killer App"?

In an SDN network, the server that is executing the control plane will also be able to run a set of additional network applications. Many of these applications are executed on individual routers in today's legacy networks. The applications will include such functionality as packet forwarding, VPN, security, bandwidth, network virtualization, load balancing, etc. However, none of these applications are important enough to cause the industry to shift to using SDN based networks.

In the computer networks that are being designed and implemented today, topology is policy. Where routers and firewalls are placed in the network will limit where broadcast domains can be, access control lists, etc.

Many enterprises are now considering moving their corporate networks from the world of the physical domain to the world of cloud computing. Having implemented a topology that allowed them to create the corporate network policy that they wanted, these enterprises would like to keep the same policy for their cloud based networks.

The problem that these enterprise network operators are now encountering is that very few of them have an abstract view of what their corporate network policy is. What they are missing is the so-called network algebra that describes who can (or cannot) talk with whom.

The true power of using a SDN based network is that the network operator can now specify a "virtual topology" of their enterprise network to the cloud. The cloud network, that has been implemented using SDN technology, can now ignore the physical design of the network that it is replacing and instead implement this policy. The end result is that the enterprise can now migrate seamlessly from / to the cloud from their current network.

In order to make this happen, the correct network policy has to be embedded into the cloud. In order to do that, the current network policy has to be "read" out of the existing network. Network operators can use their current network topology to create their network policy statement. This policy statement can then be replicated in the SDN enabled cloud. This allows things like Virtual Machines in the existing network to be moved to the cloud and the same policy requirements will apply because it will exist in the same virtual topology.

SDN technology gives enterprise network designers and operators the ability to virtualize their networks. No other technology gives them this ability and so that's why this is the "killer app" of SDN.

2 The Importance Of SDN

When we start to talk about Software Defined Networking (SDN), it is very important that we agree from the start just exactly what SDN is. It turns out that SDN can best be described as being a complete set of abstractions that allows us to specify how a network's control plane is supposed to act.

It's important to remember that in a real-world network, SDN can be implemented in a number of different ways. The wrong way to think about SDN is to view it as being a set of mechanisms – that's not what it is. When you talk with people about SDN, the subject of the OpenFlow protocol which is used to interface to the physical network's switches often comes up. You need to keep in mind that OpenFlow is actually the least interesting part of SDN from a technical viewpoint.

Unlike today's distributed network routing protocols, SDN can be thought of as simply computing a function. The SDN computes its function on an abstract view of the underlying physical network. This allows the SDN to ignore the detailed physical infrastructure that has been used to implement the actual network. What this means for network control plane behavior designers is that for the first time, they are now free of the actual physical network. The Network Operating System is then responsible for taking that function and making sure that its results get distributed to every switch in the network.

Ultimately this all leads to SDN's so-called "killer app" which is its ability to allow physical networks to be virtualized. In modern computer networks, the servers and the storage functions have already been virtualized and so virtualizing the network itself is simply the next step. With the virtualization of the network, the last stage in network designers freeing themselves from physical reality has been achieved.

Once the physical network has been virtualized, SDN then gives the software applications that are using the network the ability to reconfigure the network to suit their current needs. This ability allows the network to provide optimal service to the applications that are using the network.

3 Impact Of SDN On The Design Of Future Networks

One of the most immediate benefits that creating networks based on SDN concepts will have is that they will simplify the job of network management. However, the changes don't stop there. The wholesale adoption of SDN technology has the possibility to create very large changes to the way that networks are both designed and built.

3.1 Separation Of The Control And Data Planes

The first impact is that for the first time the network's control plane and its data plane will be able to be separated. In today's networks, the control plane and the data plane are tied closely together. The network switches that compute the routing tables are the same devices that then implement the routing tables. A side effect of this is that this means that both the control plane and the data plane are currently being provided by the same vendors.

SDN completely changes this. The control plane and the data plane are pulled apart. The control program can run on one set of servers and the Network Operating System can run on a completely different set of servers. The Network Operating system will observe and control the data plane; however, it is not part of the data plane.

One of the greatest potential impacts of implementing SDN networks is that it presents the possibility that the computer networking industry is going to be fundamentally changed. In a network that is built on SDN technology, the enterprise network designer can now purchase the control plane from 3rd party vendors and this can be done independently from the vendors the switches are purchased from. Because the "network intelligence" has now been removed from the switches and resides in the SDN layers, the network switches have been transformed into commodity hardware.

The implementation of SDN based networks will also cause fundamental changes to how network testing is performed. In today's networks, the only way to test a network is to build a "test network" with the same type of switches that you plan on deploying in the "real" network. You then hook them up in various configurations and test them to see how they perform and how well they implement the network profile that you are trying to build.

In an SDN network this all changes. Since all of the network hardware lives behind a common interface, it becomes very easy to do unit testing of the hardware in order to ensure that the interface is working correctly. Once that's done, since the control plane now exists in software, we can do large-scale simulations of it. The ability to test the network design before going live with it was one of the reasons put forth by Google when they explained why they had implemented a SDN as part of their inter-data center WAN.

3.2 The Arrival Of Edge-Orientated Networking

In an SDN network, the majority of the networking functionality that really matters can now be done at the edge of the network instead of within the network. This functionality includes such things as quality-of-service (QoS), mobility, access control, migration, network monitoring, etc. This will change what the core of the network is used for. Now the network core will be used to simply deliver network packets from edge to edge. The network protocols that are used in today's network do a very good job of doing this function.

In an SDN network, the complexity associated with the network can be pushed out to the edge of the network. This is where all of the complicated functionality associated with matching packets can be performed. Effectively what is being done is that an "overlay network is being run on the core of the network.

Pushing the network functionality to the edge of the network has two very important implications for how future SDN networks will be designed.

The first of these is that SDN will now be incrementally deployable. The network can be divided into two types of switches. The first is the core switches which will be legacy switches – the ones that are being used in the network today. The edge of the network can become software-only switches that run on a server. These switches can run a program that simulates an OpenFlow switch and this can all run on the Linux operating system that in turn is running on the Xen virtual machine hypervisor.

What this means is that SDN can be deployed without the need to make any changes to the core network. The virtual edge switches can be added to an existing network and they can tunnel across the legacy core physical switches. It is entirely possible that none of the legacy hardware switches may ever need to support the OpenFlow interface.

The other big change is that SDN offers the possibility that the network can become software orientated. The complicated parts, such as calculations around how to forward a network packet, can be done in software that runs at the edge of the network. The network's control plane is now a program that can run on almost any server. This is a big change from having the control plane be part of a closed proprietary switch / router box this is running a protocol.

How networks are created will also change. No longer will networks be designed. Rather, now they will be programmed. The focus of the "network programmers" will be on the modularity and the abstractions of the underlying physical network. No longer will the focus be on packet headers. The true value of this is that networks will now be able to innovate at software speeds and will no longer be limited to innovating at hardware speeds.

Software by its very nature lends itself to clean abstractions. What this means is that networks can now become a much more formal discipline. The clean abstractions and separations between layers in SDN will allow increased rigor.

An example of this would be how WAN control would be designed in a SDN network. One of the big challenges in this type of network has to do with link failure. When this kind of failure occurs, all of the routers in the network need to have their routing tables updated. The official way to describe this is to say that the routers need to "converge". In our example, convergence is defined to be the state of a set of routers that have the same topological information about the WAN in which they operate. To be able to say that this set of routers has converged, they must have collected all available topology about the WAN from each other via the WAN's implemented routing protocol, the information they gathered must not contradict any other router's WAN topology information in the set, and it must reflect the real state of the WAN. Another way of saying this is that in a converged network all routers must "agree" on what the WAN topology now looks like.

In an SDN network, because the Network OS will update all of the routing tables, there will be no iterative convergence for the routers in the WAN. This means that there will be a bounded depth of computation – the update will take a specified amount of time, not an indeterminate amount of time.

Looping is a big problem in today's network. Looping can occur when there is a transactional update to some of the network routing tables while a network packet is making its way across the network. The packet that was sent when the network was in State A now finds itself being routed in a network that is in State B and that's when looping can occur. Because SDN updates the abstract and global network views all at once, there will be no network disruption during convergence and so looping won't occur. This is something that can't be said about most modern routing algorithms.

One other benefit of SDN's clean abstraction is that network troubleshooting now becomes easier to do. When an operation that is not supposed to be occurring in the network is detected (packets from Node A are being routed through Node B), you can start at the control program level and check to see if this behavior is ever supposed to happen. If the answer is no, then you can go check the flow entries at each level in the abstraction. What you are looking for is the first place where they were broken.

Because the SDN control plane is implemented in software, once you've identified where the problem is, you can run a test program. By running a simulation you'll be able to identify the minimal casual set that can lead to the breaking of the rule. This means that network problems can now be detected (and solved) algorithmically.

3.3 In Conclusion

The greatest benefit of SDN will not be that data centers will be better behaved. Rather the greatest gift that SDN will provide is that it will become much easier to reason about how network control is going to behave.

When a network is built using SDN technology, it will be different from today's networks. The hardware that is used to build the network will be cheap, it will be interchangeable, and it will allow networks to follow Moore's Law. The software that will be used in SDN networks will be able to be updated frequently through software releases and the software will not be dependent on the network hardware.

Finally, the functionality of the SDN network will be driven mainly by software. This software will reside at the edge of the network in software switches and in a control program. The entire network will be built on a solid intellectual foundation.

4 An Example Of How SDN Would Work In A Network

Networks that are built using SDN technology operate differently than today's legacy networks. In order to show how a SDN network might operate, Figure 12 shows a sample network that has been constructed using four switches. Various servers are connected to each router and different parts of a given application run on different servers.

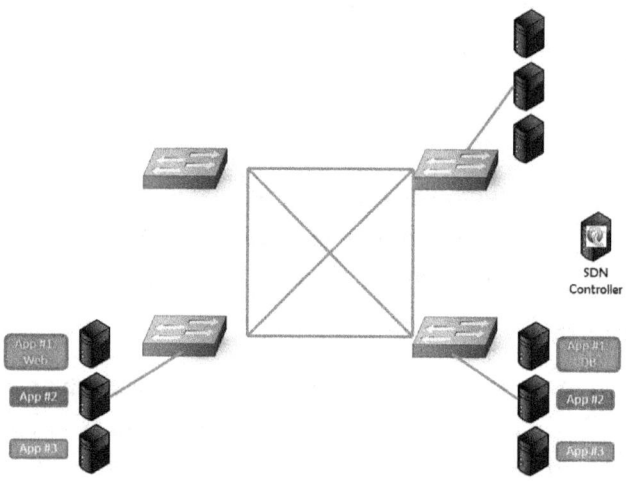

Figure 12: Sample SDN Network

The routers have been connected together to create a mesh network. In a traditional networking environment, the routers would automatically exchange information and by using the spanning tree algorithm they would quickly discover that this network contains the possibility of an infinite loop[8]. Various networking ports would be configured to be placed into blocking mode so that packet forwarding problems associated with loops would be avoided.

However, since this is a SDN network, none of those actions will occur. Instead, the SDN network controller that is shown on the right side of Figure 12 will be responsible for providing all of the routers with the data needed to populate their routing tables. Instead of concerning itself with network level details, the SDN controller will be instructed to permit the web front-end portion of App #1 to be able to talk with the database backend portion of App #1 which is located on a different server. What's actually being defined on the SDN controller are the application data flows.

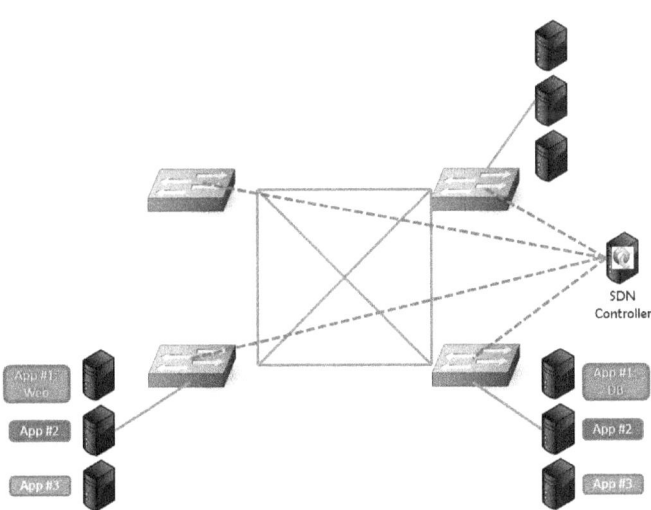

Figure 13: SDN Controller Updating Routers

Figure 13 shows the next step in the process: the SDN controller updates each of the routers in the network using the OpenFlow protocol with the flow table contents that the controller created based on the application data flows. This is unlike traditional networks where the router flow tables would have been based on such things as VLANs, spanning trees, routing protocols, subnet destination addresses, etc.

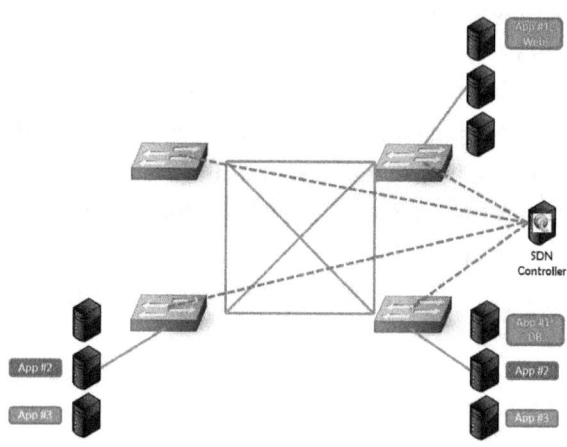

Figure 14: Application Movement In An SDN Network

Figure 14 shows an application being moved to a different server in the network. When this occurs, the SDN controller will be informed and it will then send an updated flow table contents out to each router so that the front and backends of App #1 are still able to communicate with each other.

In this SDN network configuration, the network's applications have the ability to talk directly with the SDN controller. Situations in which they might want to do this include if an application is going offline and no longer needs the network bandwidth that has been reserved for it or an ecommerce application that finds itself unable to process any additional orders and requests that orders no longer be sent to it. When the SDN controller receives these types of notifications, it will once again refresh the flow table contents on each router so that the requested network behavior is implemented.

5 How Telecom Service Providers View SDN

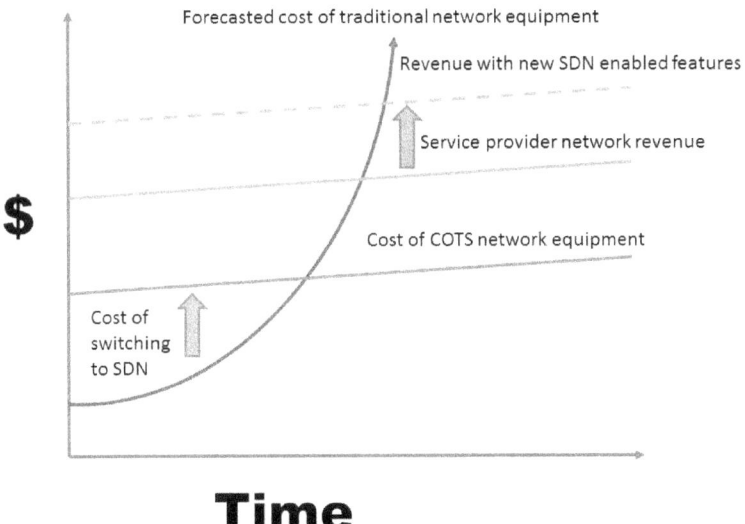

Figure 15: Telecom Service Provider's SDN Motivation

Telecom Service providers have constructed some of the largest and most complicated networks that are currently in existence. However, they are currently facing a severe problem that they desperately need to find a solution to. For these service providers, their interest in SDN is all about economics.

The cost of sophisticated networking equipment continues to increase as vendors add more and more functionality to their boxes. Unacceptably high network equipment cost escalations will result in a non-sustainable business case for the service providers. At current rates, it is entirely possible that the service provider's network service revenues will eventually be severely diminished by the cost of providing the network. Something needs to change.

Service providers have become interested in SDN because of what they believe that it can provide them with. One of the most important features that SDN can provide is a way for them to lower their cost of building and operating networks.

Service providers would like to find a way to get out of building networks out of purpose built hardware that is sold to a relatively few number of service provider customers[6]. Instead, service providers envision a network where the underlying hardware is custom-off-the-shelf (COTS) hardware that follows mass market cost curves and will result in the service providers being able to lower their equipment expenditures. The service provider's ultimate goal is to have the cost of building the network increase at a rate that most closely matches the rate at which their network revenue is increasing.

In order to make this happen, what every service provider would like to be able to do is to match their network expenses to their revenue growth. This would mean that instead of having to build an expensive network and hope that customers will show up and use it, instead they could build out the network based on the revenues generated by the customers who are currently using the network. If they could achieve this, then the service providers would be able to report consistent margins.

At the same time, service providers see a SDN network as perhaps finally providing them with the tools that they need in order to enable inexpensive network feature insertion. The belief is that these new features will result in a boost in revenues. Service providers realize that there is a risk that creation of these features may never happen and so they are not basing their SDN business cases on this occurring.

All of the benefits of switching to a SDN network have to be counterbalanced against the cost of creating a SDN network. The cost of switching from a legacy network to a SDN network can't be so great that the benefits are overwhelmed. This can be balanced out by the amount spent and how quickly the SDN network is created.

5.1 Attributes Of Successful Telecom SDN Networks

In order for a SDN network to be a successful part of the networks that telecom service providers offer service over, the network will have to have several different attributes. These attributes have been identified based on the service provider's past experience with legacy networks.

The first needed attribute is for the SDN to have a network operating system that supports a service or application orientated namespace[6]. What the telecom service providers have realized with their current networks is that they have been spending far too much time and have been purchasing too much software and hardware in order to get their networks to work at the packet level. Much of what happens in an existing network has to do with a packet's IP address and its attachment point. The service providers would prefer to be interacting with their networks in terms of service specific information such as policy routing and service routing. These service features do not necessarily tie to a packet's header information and so it can be very difficult to manage an existing network at this level.

Another key attribute of an SDN network is the ability to virtualize as many of the network resources as possible. Service providers are motivated to monetize all parts of the network that they build. This means that resource virtualization is a critical SDN need. If SDN can provide this feature, then service providers will be able to create networks that support multitenancy. This means that the network that has been built will be available for the service provider to use in order to support their services as well as those of the other service providers or enterprise users that it has sold access to. Network virtualization will provide the elasticity and the aggregation that will be necessary to allow the service provider to pool their resources in order to achieve scaling.

Service providers are looking to simplify how they interact with the network. In order to do this, they believe that the various components of the network have to be separated. This means that the topology, the traffic, and the inter-layer dependencies all have to be decoupled. In today's networks, all of these items are very closely tied together (e.g. in an IP MPLS network). Things such as IP forwarding or the services that are being provided are very closely tied to the underlying topology. The ability to separate these layers provides service providers with a powerful motivation to move to SDN based networks.

Finally, a critical attribute of a SDN network from a service provider's point-of-view is that it will support the ability to carefully introduce the new SDN architecture to interwork with the large existing base of legacy networks when the new functions will provide the greatest value. This is where potential

SDN components such as the OpenFlow control interface (along with complementary management protocols) will be used to enable new types of control paradigms on existing legacy network equipment.

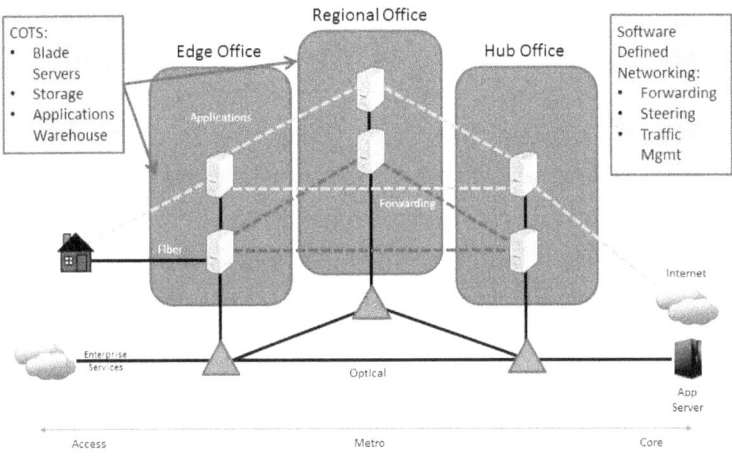

Figure 16: Deployment Scenarios For Carrier SDN

Figure 16 shows one possible deployment scenario for how carriers might start to introduce SDN technology into their networks. The left side of this picture captures the end user who could either be a consumer with access to a high-speed Internet connection or an enterprise with optical access. Note that the end users could also be accessing the network using high speed wireless technologies such as 4G LTE.

In this figure, three different types of carrier offices, edge, regional, and hub are shown. These offices are providing the end users with connectivity to either the Internet or a carrier's set of applications servers. The carrier's networks currently use their edge, regional, and hub central offices to provide telephone switching services. As SDN is introduced into the network, these offices will start to transform into data centers.

The type of equipment in these centers will start to change from specialty telephony equipment to generic servers, storage, and switching gear. Interconnecting all of these emerging data centers will be the carrier's optical transport networks.

Figure 16 shows various data flows. One such flow is the typical flow for a user to gain access to the Internet. This path will still traverse the network and will go through a gateway router. Additionally, packet forwarding is shown. This forwarding is possible to do in software if the SDN functionality to support it has been implemented. Additionally, "service aware" applications could be running and the routing of the packets for that service would be based on the type of service that is going to be provided.

This shows the vision that carriers have for introducing SDN into their networks. They are aware that this network will not happen overnight as the result of a flash cut. Rather, pieces of it will be implemented over time until eventually the complete SDN network has been put into place.

5.2 How Service Providers Plan On Introducing SDN Into Their Networks

Service providers do not view themselves as being a Research & Evaluation (R&E) lab where new networking ideas can be tried out just to see if they are good ideas. Instead, everything that a service provider does with its network has to be related to a business case and has to positively impact the company's bottom line. This being said, the service providers only plan on building functionality into a SDN external controller if that functionality can be shown to have a significant benefit for the service provider.

Examples of this kind of functionality include new feature sets or new functionality that currently is not implemented into a protocol that the service providers are using in their existing legacy networks. This can also include functions that could be done using today's protocols, but which would end up having to be implemented very inefficiently. Implementing a feature using SDN may also be done because it is believed that by doing so better scaling or economics can be realized. Finally, implementing a function in an external controller may be done if it can solve a problem that is currently not being addressed by either today's network equipment vendors or by existing network standards.

The traffic steering function provides a good example of the type of function that service providers believe may be well suited to being implanted in a SDN external controller. In this case, the service provider wants to be able to route the packets associated with a service based on that service or application and not have to deal with doing any routing based on packet headers. The routing that is done for these packets may use the IP header information as one part of a much larger data set that is used to make routing decisions. Other data that may be used could include real-time information such as the network's current congestion status or subscriber profile information. Note that much of the information that will be used to make routing decisions for the packets won't be carried in the packet itself. The software routing will be responsible for pulling a great deal of information together and then finally making routing decisions based on that. Note that this type of functionality could not be accomplished in the type of router that is being used in today's legacy networks.

The use of the OpenFlow interface may assist service providers in implementing traffic steering in an external controller. A good example of how this could work would be to look at the case in which a subscriber is watching a video over the network. Using the OpenFlow interface, the external controller could detect that a long-lived flow was being streamed over the network by inspecting the first few packets of the flow. Once this was detected, the underlying physical switches could be commanded via the OpenFlow interface to start to use cut-through switching in order to reduce the overall hardware of supporting the services that are being delivered. By reducing the load on the system that is required to support this long-lived flow, the system is better able to provide a higher level of service to other users while the video flow is being transported. This gives service providers a very efficient way to provide service aware video services.

Hybrid Cloud Computing provides another example of a situation in which service providers believe that a SDN external controller could provide them with network functionality that they can't currently get today. A hybrid cloud computing scenario occurs when an enterprise is operating their own private cloud in their own data center. When they determine that they have a need for additional computing resources for a limited period of time, it makes sense to temporarily expand their private cloud not by purchasing and installing additional servers, but rather by leasing existing cloud capacity from a service provider for a fixed amount of time.

One of the biggest challenges that enterprises face when they are looking for ways to connect their existing private cloud to a service provider's cloud is that a physical link between the two networks has to be established. What this means for the enterprise is that they are going to end up paying for bandwidth on a 24x7 basis that they won't be using most of the time. Depending on the company's specific needs, these links can by quite large in size (10G) and therefore can be quite expensive. This is one of the key reasons why service providers believe that the hybrid cloud computing service has not been adopted by customers as much as had been expected.

The SDN based solution to this problem is to virtualize the network. Once the network is virtualized, then allocation of that network can be given to the control plane which will then determine how much of the network resources the enterprise needs at any moment in time to connect its private cloud to the service provider's cloud. This then creates the possibility of offering bandwidth-on-demand as a service to enterprises who want to connect their private cloud to the service provider's cloud. The network would then determine how much bandwidth the customer needed at any point in time. This would then allow the customer to only have to pay for what they would be using.

Service providers see the lower cost of the service as making it be more appealing to more customers. The bandwidth to connect to the service provider's cloud now becomes a multitenant offering and turns into a time-sharing service. The service providers will no longer be making as much money from a single customer, but instead they'll be making money from more customers.

Another situation in which service providers believe that SDN can provide them with network functionality that they don't currently have is in the case of adding OpenFlow functionality to existing legacy routers that are operating in their networks.

Once again, there will be no "flash cut" in which all of the existing routers will be removed from the network only to be replaced by SDN routers. This means that the service providers are looking for a graceful way to move from today's network to a SDN network in the future. Adding OpenFlow functionality to existing legacy routers is one way to go about doing that.

Adding OpenFlow functionality to a legacy router can be done in several different ways. One way is the scenario where a router already has its native control that has been provided by its manufacturer. To this configuration, OpenFlow can be added in order to create a hybrid mode where both control sets exist within a single router. Each set of interfaces would operate separately; however, there would have to be a system established to allow them to share the routers physical resources.

Finally, service providers are very concerned about how SDN networks are going to scale. Today's service providers operate very large networks and they realize that tomorrow's SDN networks will be just as large, if not larger. This means that they need to find solutions to issues having to do with OpenFlow switch partitioning and the best way to support multiple SDN controllers.

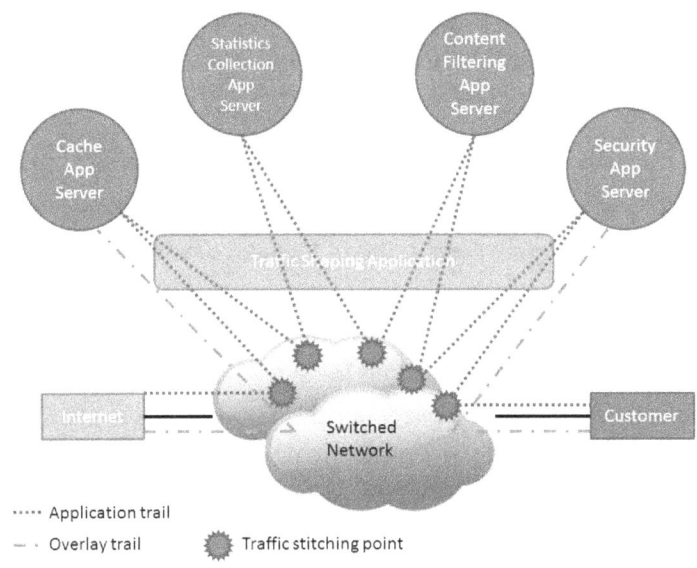

Figure 17: How Traffic Steering Is Done Today

Figure 17 shows an example of how traffic switching is being implemented in a service provider's network today. The traffic arrives in the network and it then routed from the traffic shaping application to a series of applications. Initially, the packets will follow the application trail shown in the figure as the network discovers what the communication session is all about. Once that is known, there may no longer be a need for all of the packets to be processed by all of the applications. Once that happens, the remaining packets can follow the overlay trail and only be processed by the applications that truly need to see each and every packet.

The application "traffic stitching points" exist in the traffic shaping algorithm between the overlay trails. The purpose of this is to allow the development of arbitrary network feature graphs that can be allowed to vary over time. It is the purpose of the traffic steering algorithms to determine the application trail that will be taken through both the service features and the cache.

This implementation of traffic steering does provide both flexibility and can be considered to be extensible. However, it requires network packets to travel through a large number of network interfaces and requires a great deal of processing power to be spent processing each and every packet. This processing may not be required for every packet and when long-lived flows are traveling over the network this is a less than optimal solution that will use up too many network resources that could be better used processing other flows.

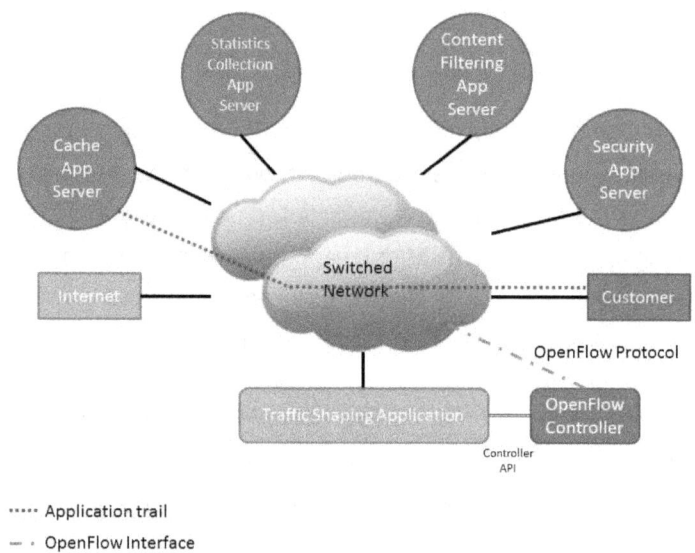

Figure 18: Using OpenFlow For Traffic Steering Optimization

Figure 18 shows how a traffic steering application might be implemented in a SDN network. One of the goals of this network design is to allow network packets to avoid being processed by the traffic shaping application if that type of processing is not needed. In the example shown in Figure 18, the customer is retrieving data from the network's cache application server after the network has processed this data for security and content filtering. Once the network determines that it will be processing a long-lived flow, it can reduce the amount of processing that it does on each packet in the flow by starting to process them using cut-through routing.

Because network statistics are still important even for this type of flow that will not be processed by the traffic shaping application, the statistics for the long-lived flow will be collected by the OpenFlow interface. Another way of implementing this would be to have the OpenFlow controller determine flow usage for a particular pattern and use this information to determine when a flow completes.

In Figure 18, The OpenFlow Controller is shown as being a separate application. The OpenFlow controller API could be implemented as a proprietary API. Alternatively, this software could be implemented as a part of the traffic shaping application.

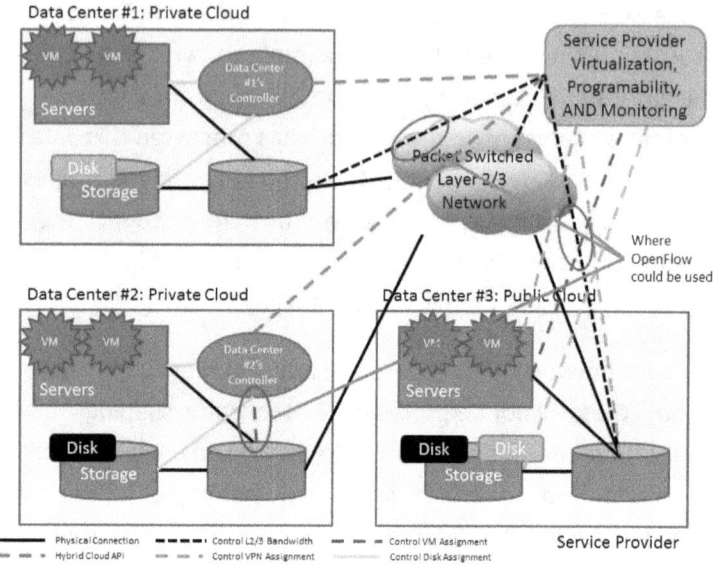

Figure 19: Example of Bandwidth-On-Demand For a Hybrid Cloud Architecture

Figure 19 shows an example of a network that could be used to provide an enterprise with access to a service provider's hybrid cloud resources. In this scenario, OpenFlow could be used to organize how the data would be routed through the routers.

A key part of this figure are the hybrid cloud APIs which do not currently exists. However, if the idea of a hybrid cloud is going to gain acceptance for enterprise customers, this part of the SDN technology is going to have to be developed by service providers and equipment providers in order to permit seamless integration between enterprise private clouds and service provider clouds. These APIs will allow multiple users to make use of the underlying network resources. This will determine how resources are requested and then controlled once they are granted.

5.3 Summary

Service providers have become very interested in using Software Defined Networking (SDN) to create the next generation of service provider networks. Their motivations are driven by the practical underlying economic realities of today's legacy networks: they are becoming too expensive to both build and maintain. The increasing cost of the network equipment that is being purchased and deployed in today's networks coupled with the larger and larger volumes of equipment that the growing networks are requiring are driving service providers to look for more economical ways to build their networks.

A network built using SDN technologies offers service providers the possibility of finally being able to get their network costs under control. Building a network based on a COTS infrastructure offers the service provider a way to align their network's cost structure with their service revenue.

Implementing a SDN network will fundamentally change the way that a service provider designs and deploys their networks. The Central Offices that are a part of today's legacy service provider network will be transformed from holders of special purpose telecommunications equipment into data centers. This will allow the service providers to be able to reap the benefits that come along with the cost, scaling, and new service flexibility that is offered by the arrival of the technologies associated with cloud computing.

The use of SDN technology to build the network will allow the network to become aware of the types of traffic that is traversing it. This means that it will now be possible for some types of network traffic (e.g. video traffic) to be handled initially by SDN and then by cut-through switching. This means that fewer network resources will be used to transport these packages and therefore will be made available to support other network services and traffic.

The use of SDN technologies opens the door for fuller enterprise adoption of hybrid cloud computing. When implemented using SDN, hybrid cloud computing can use the OpenFlow interface along with new ways of managing network traffic in order to create a seamless interface between the enterprise private cloud and the service provider's cloud. New hybrid cloud APIs will need to be created and adopted by the industry in order to fully enable this new service offering.

Of the greatest concern to service providers is how SDN technologies can be made to work with existing legacy networks. Due to their heavy investment in today's networks, service providers are very interested in finding a clear path from where their networks are to the fully SDN enabled network of tomorrow. In order to make this happen, several changes in terms of virtualization and resiliency will have to be made to the OpenFlow interface in order to ensure that it can coexist with today's legacy switching systems.

6   The Developer And The Network

The creation of networks based on SDN technology holds out the promise of networks that can be programmed to better support the traffic that they are being asked to carry. Although the possibility of implementing SDN technology in order to make networks more cost efficient is attractive, creating networks that can be tuned to the traffic that they are carrying is even more exciting.

6.1 The Problem With Today's Applications And Networks

The two major problems with today's networks are that the applications that are using the networks don't know enough about the underlying network that they are using and the network doesn't know enough about the applications that are using it. A wide variety of different techniques have been created in order to attempt to solve this problem; however, none of them do a very good job.

Current approximation techniques are barely sufficient and are basically ineffective. Applications are forced to guess about the networks that they will be using to carry their packets. In order to make these guesses, applications have to rely on network tools such as ping-stats, Doppler, geo-location, and whois lookups. The network equipment that the applications may try to talk with are all using different forms of proprietary codecs which require custom proprietary interfaces and the best that any application is ever going to do is to be able to create an approximate network topology and guestimate where their location within this network is. A much better way to go about getting the network information that an application needs would be for the application to be able to ask the network.

Things are no better on the network side. Today's networks have to resort to spying on the application traffic that they are transporting in the hopes of being able to gain a better understanding of the applications that are using the network. Networks have their own special set of tools that they use to do this: deep packet inspection, stateful flow analysis, application fingerprinting, and service specific overlay technologies. All of this effort is being expended in order to allow the network to try to maintain the service level agreements (SLAs) that have been promised to the applications that are using the network. All of this turns out to be very expensive in terms of the network resources that it consumes.

One way to look at today's network is to view it as being an interaction between four separate entities. These entities would be applications, application data, users, and the network. Each one of these entities has a different view of what is going on:

- **Applications**: Applications are responsible for knowing what the capabilities of the end user devices that have been attached to the network are because this is how the user will be interacting with the application. Additionally, how close the end user is to the application's content is also critically important especially for real-time applications such as games. Ultimately, the application will be responsible for controlling the network resources.

- **Content**: The idea that content can become network aware changes how the content will interact with the network. It will be the responsibility of the content to make modifications to its placement within the network along with controlling how content gets selected. The actual insertion of content into network data flows can be controlled by network based analytics.

- **User**: Perhaps the least complex of all four components. The end user is aware of what he wants from the application and therefore from the network. In order to get his desired information, the application will be responsible for directing the user where to go on the network in order to get it.

- **Network**: The network lies at the center of the interaction between all four entities. This means that it is responsible for the real-time interaction between the user, the content, and the application.

In order to enable the interaction between these four entities, a new form of bidirectional communication needs to be implemented in the network. The applications have to be able to talk to the network and the network has to be able to talk to the applications.

6.2 What Is A Programmable Network?

One of the most promising features of SDN technology is that it opens the door to permitting the network to be programmed. It's important that exactly what this means be explored. The key functionality that a programmable network provides is the ability of the applications that are using the network to be able to inform the network about what the desired network behavior that they need is. At the same time, the network needs to be able to inform the various applications that are using it about the data that is intrinsically contained within the network.

The way to permit this type of application / network communication to occur is to establish new so called network "touchpoints"[7]. These programmable touchpoints are what will actually be programmed in order to create the network behavior that will be needed in order to support the applications.

The types of touchpoints that will be supported can be divided into two high-level groups. The first is user based and the second is network based. User based programmable touchpoints include such data sets as a user service profile, a billing profile, and a security gateway that could include both VPN information and mobile information.

The network programmable touchpoints are more focused on the behavior of the equipment that is being used to implement the network. These touchpoints would include such items as the enterprise edge, business edge server profiles, content delivery networks (CDN), and even hypervisor stack information in order to allow the management of how virtual machines (VMs) interact with the network.

6.3 The Value Of Having More Network / Application Information

In a SDN network, applications will have access to information about the network that they are using that is either not currently available to them or is difficult to tease out of the network. This new type of information will include information about the end user devices that are attached to the network and are accessing the applications.

Applications can be informed about where they are located in the network's topology – the real location not a guess, what access network technology they are using, how much bandwidth is available, and the utilization of the link that they are using to connect to their endpoints. With these new types of network information, applications will now have the ability to adjust their behavior to better match the network's real-time usage. These types of adjustments may have an impact on how much the user is willing to pay for the service that they are receiving so additional billing granularity will then also have to be supported. A final benefit of having greater access to more network information is that applications will then have more flexibility as to where they are physically placed in the network.

Networks can become more efficient with the information that applications will be able to provide to them in an SDN network. Ultimately, it will be the applications that end up controlling the network resources. In order to do this, the network will have to use application provided data to optimize both network bandwidth and network resources. This will allow new service topologies to be defined and supported. Security identification will be able to be included into the transport of all data across the network. Finally, with application provided information, networks will finally be able to provide service-specific packet treatments.

6.4 Evolving Legacy Networks To SDN Networks

Today's legacy networks will not transform into SDN based networks overnight. How that evolution will occur is a topic of great debate in the networking community. There seems to be mutual agreement that it will occur by having the SDN technology augment what is already on the network. All players agree that SDN must not break the networks that exist today!

As SDN enabled equipment starts to be introduced into the network, there will have to be integration with the existing routing, signaling, and network policy logic. The true value of

SDN is that it makes the network programmable. This means that it will need to support programmable touchpoints that are modular in design.

The service model that will be presented to users must be seamless. This can be accomplished by creating collaborative inputs. Ultimately, in order for the various service provider and vendor players to get involved, SDN products are going to have to be standards based.

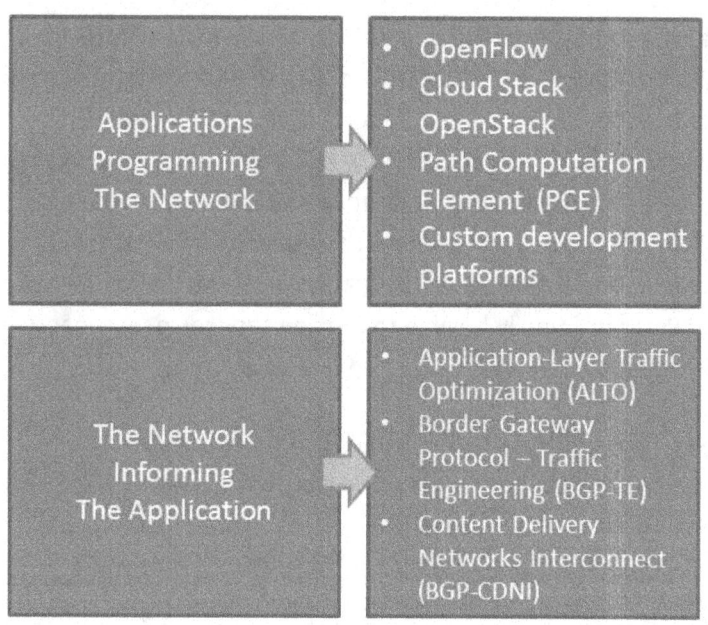

Figure 20: How applications and the network will interact

Figure 20 shows how applications and the underlying network could communicate in a SDN network. The most important point to realize here is that the needed information will not be exchanged using a single protocol that has been loaded down with all of the API information that is needed to support every application and every network. Rather, a collection of protocols can be used to provide the right application or the right network with the information that it needs. It's foreseen that a SDN will use a collection of protocols to provide a modular workflow approach in order to get information out of the network as well as to program it into the network.

Providing this set of protocols will solve a number of the problems that exist in today's networks. Routers or switches in a legacy network do not currently have an API (or even a Software Development Kit [SDK}) that allows them to program an Access Control List (ACL). This functionality currently has to be performed by a network administrator using a command line interface (CLI) or the Simple Network Management Protocol (SNMP).

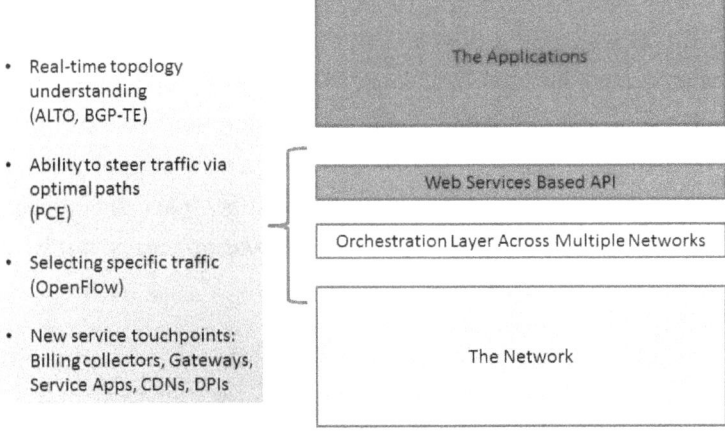

Figure 21: Proposal For Adding SDN Without Breaking The Internet

Figure 21 shows one way that SDN could be made to work with existing legacy networks. In this example, the complicated network protocols and interfaces that are currently used are encapsulated in the network. On top of this an orchestration layer is created that will provide an interface between today's collection of network protocols and a higher level web services based API. This is the API that application developers would then develop their software to interface with. This abstraction would allow the application developers to not have to deal with the technical details of the networking specific protocols.

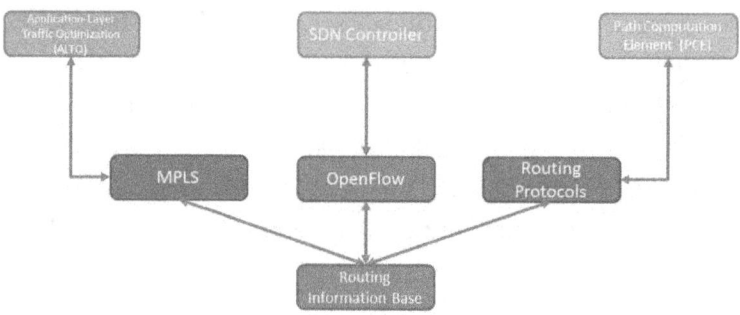

Figure 22: Adding SDN To A Legacy Router

Figure 22 shows how SDN and OpenFlow could be added to a legacy controller's control plane. In this figure, support for the OpenFlow protocol is being added to augment what is currently available in routers today. This figure shows the support for the OpenFlow protocol existing in parallel to the legacy protocols. This architecture has been referred to in various ways, but is most commonly called the "ships in the night" approach.

The ships in the night approach to adding OpenFlow support to a legacy router would require that the router's physical resources be partitioned. A portion of these physical resources would be given to the OpenFlow functionality to control along with a portion of the router's ports. Another portion and a matching set of ports would then be given to the router's existing native control program to control as it does today. It is anticipated that there would be some level of integration between the OpenFlow and the native control program in order to ensure the proper operation of the router.

An alternative way of adding OpenFlow support to a router is to integrate the support for OpenFlow with the controller's existing legacy protocol logic. In this scenario the OpenFlow functionality could be used to define features. This would allow the native control plane to be augmented with additional functionality. The integration of both types of control plane logic means that the routers resources would no longer have to be partitioned. This solution would allow the router to operate at several different abstraction levels depending on the type of network services that were being supported.

The current belief in the networking community [7} is that both types of OpenFlow implementations will eventually be supported in an SDN network. The belief is that there is room in the market for both types of solutions and the different price points and functionality that they will bring to the market.

7 Examples Of Networked Applications That Can Only Be Offered In An SDN Network

The arrival of SDN technology will open the door to a whole new set of applications that can now be developed. Functionality that is not currently available in today's networks will then be able to be deployed in tomorrow's SDN networks. At this point in time we can't be certain what types of new applications will be developed; however, we can take a guess at a few of the new types of applications that we will be seeing.

7.1 Service Engineered Paths

Today's traffic engineering paths are set up using protocols such as RSVP. When we program classifiers in front of them using OpenFlow, we are creating service engineered paths. We have now identified the specific service that we want out of these flows and we have put them into a service topology or guaranteed the reserved bandwidth over that network and this can then be used to provide a new set of services.

A tunneling or switching technology is used to provide a path that is then used by specific functions of a given service. This will then allow the network to perform selective traffic redirection based on transient classifiers. These signaled paths would be set up using the Path Computational Element (PCE) standardized API.

7.2 Service Appliance Pooling

For cost savings, you can identify specific traffic and then program the classifiers into the forwarding plane and take the (on average) 15-20 service appliances that exist around an edge router today and pool them back into a data center. Once this is done, these service appliances can then be used to support multiple edge routers.

7.3 Content Request Routing

Content request routing has to do with identifying where the end user is located and then identifying where the data that they are trying to access is located. The user provided information will use information based on the network infrastructure. This will need to be able to execute across multiple service provider networks and it will have to support both mobile and broadband users.

The determination of which copy of the data should be used to support the end user will be based on a number of different factors. These will include network proximity, network availability, network congestion, the availability of the content, how much of a load the content will place on the network, and the capacity of the content. This type of information can be provided by the new ALTO (Application Layer Traffic Optimization) protocol.

The goal is to allow the network to identify the best location to use for accessing the requested content. Within an SDN network, this answer can be provided by the network's topology and once the answer is known, it can be provided to the DNS (Domain Name System) server so that end user requests are properly resolved.

7.4 Bandwidth Calendaring

Within today's networks, there are pre-planned activities that require a fixed amount of network bandwidth in order to support them. These types of activities can include adding flexibility to where services are placed within the network, the ability to schedule when data center content is backed up, management of the distribution of application content, and the orchestration of cloud related activities.

Bandwidth calendaring refers to the ability to schedule the bandwidth that an activity is going to require. This means that a specific network path will be reserved in order to support your application. The use of a SDN means that this path can be made available to your application when it is needed and, most importantly, it means the bandwidth that you need won't be taken away from you when another high priority application starts up in the middle of your session. Only your traffic will be permitted to travel over your reserved network path.

In order to implement bandwidth calendaring, four separate technologies will be required to be used. The first is the use of the ALTo and BGP-TE protocols in order to provide a real-time understanding of the current network topology. Next, the PCE protocol will be used to steer the application traffic through the optional network paths. The initial reservation will be established using a Web Services API. Finally, the specific traffic that will travel over the reserved path will be selected using the OpenFlow API.

7.5 Social Networking

In the world of social networking, for a wide variety of reasons, where the end user is located is very important. However, the technologies that are being used in today's legacy networks can at best approximate the user's location. There is additionally a great deal of useful information that is currently not available to social media applications. This "missing information" includes the access technology that the user is using and the bandwidth that is currently available to them, the capabilities of both the device that they are currently using and the network connection that they have connected though, their specific geographical location, how close they are to network content, how much bandwidth they have paid for and any security related issues.

Today's networks attempt to determine where a user is located by using three different methods. The first is an active broadcast by the user of where they are located – effectively "checking in". The next is a so-called "game broadcast" which occurs when you move to a new access link. Finally, there is the passive derivation where the network uses what information is available to try to determine the end user's location.

The goal in an SDN enabled network is to allow social networking applications to provide a continuous real-time streaming of their resources, their people that they are interacting with, their location, and their surrounding network content. The end user will be defined by data that includes their access, capacity, bandwidth, and their profile. Additionally, because their location will be known, additional information will be available including content, resources, places, and people.

8   How Vendors Are Responding To SDN

The arrival of the Software Defined Networking approach to building networks brings with it the possibility of completely changing how both carriers and enterprises build their networks in the future. This change could have very significant impacts on the companies that provide firms with the network hardware that they use to build their networks.

The network hardware that is being used to build networks today is very sophisticated. Once a firm selects a specific vendor's equipment, they are committed to buying more equipment from that vendor as their network grows along with the necessary software and hardware upgrades that occur over time. SDN may change all of this.

In a SDN network, simple high-speed routers are controlled by sophisticated software applications that run on separate servers. This architecture is completely different from what is deployed in networks today. In order to be able to survive and grow in this new networking environment, all of today's vendors are going to have to change significantly. This is an evolving story; however, as of the time of the writing of this book, here were the initial steps that each of the major vendors were currently taking.

8.1   Cisco's Response To SDN

Today's legacy networks have been built using sophisticated routers that are provided by networking equipment companies. Cisco is one of the largest and most successful of these companies. Clearly, the arrival of SDN and its use of simple packet forwarding hardware poses a significant threat to Cisco's primary source of revenue. How Cisco responds to the threat posed by SDN will have a significant impact on how networking is done in both the short and long term.

In November of 2013 Cisco announced their initial response to SDN: the Application-Centric Infrastructure (ACI). Cisco realized that they needed additional expertise to create a strategy to deal with the threat posed by SDN and so they acquired a company called Insieme Networks.

Cisco had already held an 85% ownership stake in Insieme and they had made repeated investments in the company, including US$100 million in April 2012 and an additional US$35 million in November 2012. The end result of these investments was that in November of 2013 Cisco announced that they were going to fully acquire the software-defined-networking-focused firm.

The products that Insieme was developing were the Nexus 9000 line of data center and cloud switches that featured application awareness in order to make the network infrastructure flexible and agile and permit a dynamic response to application needs and virtual machine workload mobility.

The Nexus 9000 is Cisco's hardware-based response to SDN. The Nexus 9000 switches contain both "off the shelf" chip components ("merchant silicon") and custom silicon for both basic network virtualization (OpenFlow and VXLAN) and Cisco proprietary application-centric networking, which is being called the "Application Centric Infrastructure" (ACI). The big difference between Cisco's approach and the open source SDN approach is that Cisco wants the network to be application aware vs. the applications to be network aware. ACI is hardware designed to equally provision both physical and virtual resources in data centers and cloud networks no matter what hardware or hypervisor those resources are based on.

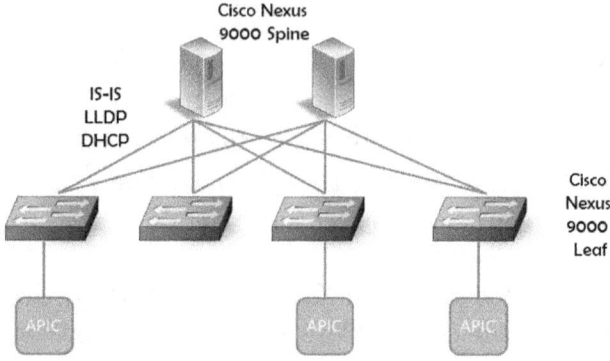

Figure 23: Cisco ACI Fabric

The Cisco ACI fabric consists of the three components shown in Figure 23: the Cisco Application Policy Infrastructure Controller (APIC) and Cisco Nexus 9000 Series leaf and spine switches. In an ACI network the leaf top-of-rack (TOR) switches attach to the spine switches, and never to each other. The spine switches attach only to the leaf switches, and possibly to a higher-level spine switch if the network design is hierarchical. The Cisco APIC (and potentially all other devices in the data center) attach to the leaf switches only.

In an ACI network, services are supported by providing a REST API that can be used for automation. ACI provides a service graph that is composed of logical service functions that are communicating to the outside using endpoint groups. The ACI service graph characteristics include: traffic filtering based on policy, taps, graph splitting and joining, traffic reclassification, etc. ACI has the ability to provide resource pooling for stateless load distribution across multiple destinations, each of which can then perform its own stateful load balancing. Simple pooling is supported by devices that are unaware of ACI and more advanced pooling can be provided by those that are aware.

Cisco's Application Policy Infrastructure Controller (APIC) provides ACI with the brains to implement both policy and management[10]. APIC is a critical part of Cisco's software-defined networking (SDN) plan. APIC does the following functions: assigning policy to a traffic flow, having the policy move with it, creating service chains to apply the policy properly, and automatically attaching policy to certain types of workloads.

The APIC is a cluster of controllers that have the following characteristics: they are distributed, they offer a single point of control, provide a central API, and they contain a repository of global and policy data. APIC Policies are distributed under a variety of conditions. An example of this would be "just in time" when a node attaches, or statically. Node attachment is detected by the APIC using triggers. Cisco has stated that the scalability of APIC includes 1 million plus endpoints, 200K+ ports, and 64K+ tenants [10].

The Cisco Nexus 9000 router will also support the OpenFlow interface and the OpenDaylight open source controller, but only in "off the shelf" chip component -based "standalone" mode on the Nexus 9000. Full function ACI mode can only be achieved using Insieme's custom ASICs in the Nexus 9000.

Not only does ACI require new switches (the Nexus 9000 line), but they're not line-card compatible with the old switches (the Nexus 7000s), and they will require a software upgrade ACI comes around.

The OpenDaylight controller does not play any role in Cisco's design[9]. Insieme developed its own controller for ACI-mode networking, which is achieved by using Insieme's custom ASICs. In "standalone" mode, the Nexus 9000 with "off the shelf" silicon – in this case, Broadcom's Trident II – can support OpenDaylight, OpenFlow and other open source software for SDNs, and network programmability and virtualization.

Even though it is easy to criticize Cisco's monolithic architectures, it is entirely possible that some customers will prefer it to other alternatives. One way of putting it is that Cisco is offering their customers "one throat to choke" – a single vendor who provides all the pieces of the SDN solution.

Initial feedback from carrier customers regarding Cisco's SDN plans has not been favorable [12]. The U.S. telecommunications provider AT&T released their **Supplier Domain Program 2.0** Request for Information (RFI) at the end of 2013. In this document, AT&T asked their vendors, Cisco included, to explain if they are developing SDN controllers and how their network equipment will be adapted in order to be controlled within a SDN network.

It has become clear that AT&T is interested in shifting away from purpose-built smarter hardware and toward commercial off-the-shelf boxes as part of their move to using software defined networking (SDN) and network functions virtualization (NFV) to save significant network capex expenses.

Market research firms believe that AT&T is not going to be open to Cisco's Application Center Infrastructure (ACI) architecture, which includes the Nexus 9000 switches, because it still seems too complex and proprietary compared to more white box-oriented architectures.

What is going to be most important for Cisco is if AT&T decides to move completely away from ASIC-based hardware with some intelligence and toward a "white box" approach that is being favored by software players such as VMWare.

8.2 VMware's NSX SDN

VMware started the development of its SDN solution early with the US$1.2 billion acquisition of startup Nicira in mid- 2012. The motivation for this purchase was Nicira's network virtualization strategy. It fit well into VMware's overall product set, allowing for a tight coupling with products such as vSphere. Just over a year after the Nicira acquisition, VMware announced its network virtualization platform called NSX that had been created using Nicira technology in August 2013.

VMware, the leader in server virtualization, is using its NSX product to branch into network virtualization. NSX is a software-based network virtualization overlay that enables a VMware hypervisor to provide network control functionality as shown in Figure 24. It is the direct opposite of what Cisco proposes and should be viewed as a direct competitor to the Cisco / Insieme's ACI and Nexus 9000 approach.

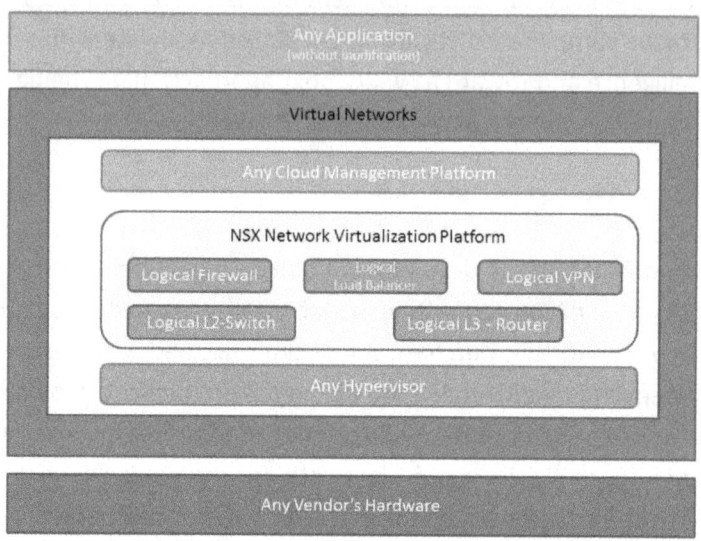

Figure 24: VMWare's NSX Virtualization Platform

NSX provides network virtualization by provisioning hypervisor virtual switches to meet an application's connectivity and security needs. NSX uses virtual switches which are connected to each other across the physical network using an overlay network. How NSX accomplishes this revolves around a distributed virtual switch (the VMWare vSwitch product).

The vSwitch sits at the network edge in the hypervisor and handles links between local virtual machines. The vSwitch provides access to the physical network If a connection to a remote resource is required. [24]

The NSX controller arbitrates applications and the network. The NSX controller uses northbound APIs to talk to applications, which communicate their needs. The NSX controller then programs all of the vSwitches under NSX control in a southbound direction to meet those needs. The NSX controller can use the OpenFlow protocol for those southbound links, but OpenFlow is not the only part of the solution, or even a key one. In fact, VMware de-emphasizes OpenFlow in general. [24]

VMware NSX creates a virtual network overlay that is loosely coupled to the physical network underneath. Cisco and other vendors argue this approach won't scale well. Network experts believe that whether software switches and virtual network overlays are enough to handle high-performance environments depends on the networking situation.

It is agreed that the in the cases where a simple software switch is performing a fairly basic Layer 2 feature set, the vSwitches (software switches) in ESX can do a great job today. Other cases such as a software switch performing more advanced services in an open source hypervisor will not be able to provide practically jitter-free, sub-millisecond performance. In this case removing the hypervisor from the network path and replacing it with dedicated network hardware still makes sense. [25]

8.3 Juniper

Juniper Networks purchased Contrail Systems, a startup maker of software-defined networking (SDN) software, in December 2012 for US$176m. Contrail's goal had been to create a controller that would be compatible with the OpenFlow protocols that had come out of Stanford University. However, the Contrail products were based on existing network protocols and would therefore be compatible with existing switches, routers, and server virtualization hypervisors.

The Juniper controller is based on the Border Gateway Protocol (BGP) that is already embedded in Juniper switches and routers. The controller also employs XMPP, a protocol for transmitting message-oriented middleware messages, to control the virtual switches inside of hypervisors. Juniper has decided to use an existing technology from telecom networks called Multiprotocol Label Switching (MPLS), which encapsulates packets on a network and controls their forwarding through those labels; MLPS exists between Layers 2 and 3 in the network stack. [26]

There are several parts to Juniper's Contrail platform as shown in Figure 25. The first component is a software controller. This software controller has been designed to run on a virtual machine and supports a redundant active-active cluster configuration. Cloud tools can interact with the controller using a set of RESTful APIs that have been implemented in order to support northbound interaction. When it was launched, Juniper's Contrail platform had been certified to interwork with OpenStack and CloudStack, as well as with IBM's SmartCloud Orchestrator. [27]

The second component of the Juniper controller solution is the vRouter which can be run on either the KVM or Xen hypervisors. The Contrail controller will communicate with the vRouter using the XMPP protocol and this will permit it to tell the vRouters how to forward network packets. The vRouters will be responsible for building tunnels that run over the physical network between virtual machines that are active in the network.

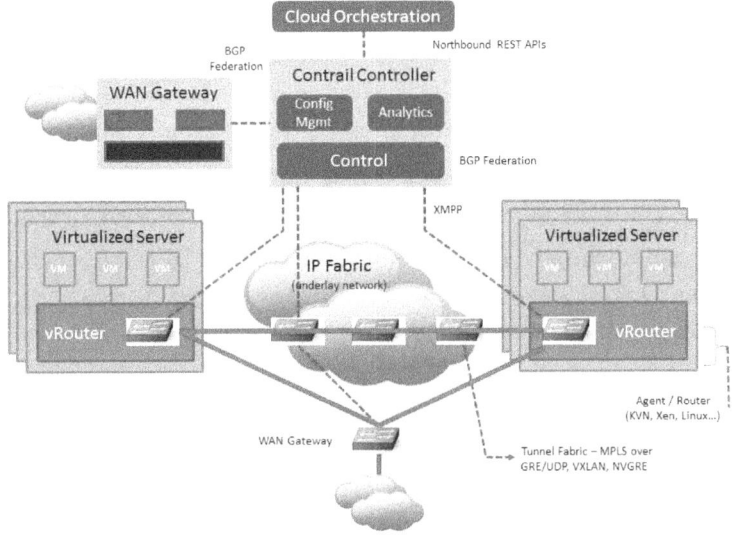

Figure 25: Juniper's Contrail Controller

The initial release of the Juniper controller will not provide support for the OpenFlow protocol. The plan is to adopt a wait-and-see attitude in order to determine how the market reacts to OpenFlow.

One of the biggest differences between an OpenFlow-based controller and Juniper's controller is that the Juniper solution keeps the master copy of the forwarding tables on the controller and copies them out to the switches rather than keeping the master copies on the switches and aggregating them on the controller after they have been changed.

Juniper, which is a member of the open source controller project OpenDayLight has announced that they will be making the code for their Contrail controller open source. Why they have decided to do this instead of contributing it to the OpenDayLight project is unknown at this time.

8.4 OpenDaylight SDN

The OpenDaylight project is an open source project with a modular, pluggable, and flexible SDN controller platform at its core. This controller is implemented strictly in software and is contained within its own Java Virtual Machine (JVM). As such, it can be deployed on any hardware and operating system platform that supports Java.

The OpenDaylight project is currently supported by 31 networking industry companies. These supporters include Cisco, IBM, Juniper, Microsoft, Redhat, VMWare, Ciena, Intel, Dell, HP, and many more.

The OpenDaylight community is developing an SDN architecture that supports a wide range of protocols and can rapidly evolve in the direction SDN goes, not based on any one vendor's purposes.

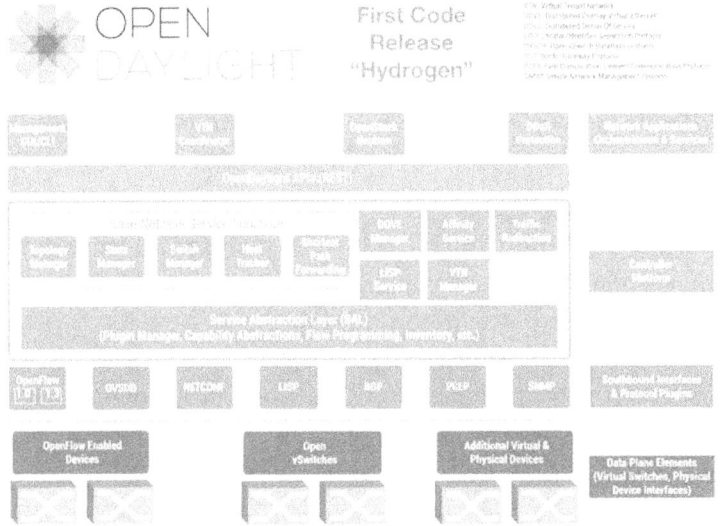

Figure 26: Technical Overview Of The OpenDaylight Controller Hydrogen Release [Source: OpenDaylight project]

The OpenDaylight controller provides open northbound APIs which are designed to be used by applications. The OpenDaylight controller supports the Open Specifications Group Initiative (OSGi) framework and bidirectional REpresentational State Transfer (REST) for the northbound API.

The OSGi framework is used for applications that will run in the same address space as the controller while the REST (web based) API is used for applications that do not run in the same address space (or even perhaps on the same machine) as the controller. In the OpenDaylight controller, the business logic and algorithms reside in the applications. In order to perform their functions, these applications use the controller to gather network intelligence, run algorithms to perform analytics, and then use the controller to propagate the new rules to network devices, if any, throughout the network.

The OpenDaylight controller platform contains a collection of dynamically pluggable modules that perform needed network tasks. There are a series of base network services for tasks such as understanding what devices are contained within the network and the capabilities of each, statistics gathering, etc. Additionally, platform oriented services and other extensions can also be inserted into the controller platform for in order to provide additional SDN functionality.

The OpenDaylight controller's southbound interface has the ability to support multiple protocols. These protocols can be added as separate plugins. Examples of protocols that can be added include OpenFlow 1.0, OpenFlow 1.3, BGP-LS, etc. These plugin modules are dynamically linked into a Service Abstraction Layer (SAL) of the controller. The SAL interfaces device services to which the modules north of it are written. The SAL determines how to fulfill the requested service independent of the underlying protocol used between the controller and the network devices.

OpenDaylight's first release of their open-source controller, named Hydrogen, is shown in Figure 26. This controller will include new and legacy protocols such as Open vSwitch Database Management Protocol (OVSDB), OpenFlow 1.3.0, Border Gateway Protocol (BGP) and Path Computation Element Protocol (PCEP). [28]

Hydrogen will also include multiple methods for network virtualization and two initial applications that leverage the features of OpenDaylight: Affinity Metadata Service to aid in policy management and Defense4All for Distributed Denial of Service (DDoS) attack protection.

To make OpenDaylight more cloud friendly, it will include an OpenStack Neutron, OpenStack's virtual networking, plugin and the Open vSwitch Database project will allow management from within OpenStack.

8.5 Big Switch Networks

Big Switch Networks is a well-funded network virtualization and SDN company that provides several products based on the OpenFlow protocol. Big Switch Networks has created three main products:

1. **Big Network Controller** : This controller is a commercial version of the open source Floodlight controller. The controller is a platform on top of which software applications run. This automats the underlying fabric and allows an entire network to be controlled from a single console.

2. **Big Virtual Switch**: This component virtualizes the network using existing servers, enhancing network flexibility and making use of resources in the most efficient manner possible.

3. **Big Tap**: This is a network monitoring application that uses OpenFlow-enabled switches to provide network administrators with full network visibility. This permits them to scale the network in order to minimize operating costs.

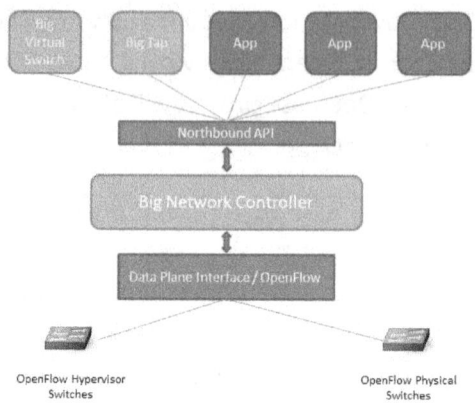

Figure 27: The Big Switch Networks Open SDN

Big Switch's Floodlight-based controller, shown in Figure 27, initially only talked to Open vSwitch switches and the unnamed virtual switch inside of Hyper-V. It could support up to 250,000 new host connections per second and talk to over 1,000 physical or virtual switches on a single two-socket x86 server. [29]

9 Google And SDN

Google is one of the most successful Internet companies in the past few years. Their search engine technology and advertising products have provided them with immense financial success. What this means is that the networks and the data centers that they use as a part of their business are critical to the company's success. Additionally, they have deep financial resources and can invest in almost any networking technology that they want to in order to optimize how their networks perform.

This is what it was such a significant event when in April of 2012, Urs Hölzle gave a presentation at the Open Networking

Summit in Santa Clara, California [5]. During this presentation Urs stunned his audience by revealing that Google had built and was using a SDN based network to interconnect its data centers. Note that because of when they had done this, there were virtually no vendors or products available for them to purchase – they had to create everything that they used by themselves.

During his presentation, Urs revealed that Google's network contains two separate large backbone networks. One of these backbone networks has been designed to be Internet facing and is the one that carries user traffic. The other backbone network has been designed to carry internal Google traffic and is used to interconnect Google's datacenters.

9.1 Google's Problem With Its Network

Google's experiences with running its own network has shown it that running and managing networks is not an easy task – in fact it is quite hard. In order to simplify this complex task, Google has started to use OpenFlow in order to accomplish two tasks. The first is to improve the performance of their backbone network and the second is to reduce the complexity (and associated cost) of managing the backbone.

Google's networking needs are very large. Measurements taken in 2010 revealed that Google was responsible for 6% of the entire Internet's traffic. If Google was an Internet Service Provider (ISP), they would be the second largest ISP in the world. Many Google products, such as YouTube, web search, Google+, Maps, and updates to the Android operating system and the Chrome browser all generate a great deal of network traffic.

Urs stated that Google had run into a significant problem. In a data center environment, in general as the scale of an application increases from a single server to multiple servers, the cost of running that application as measured in CPUs and storage decreases. However, the problem that Google was having was that this same effect was not being realized when it came to networking. The cost / bit was actually increasing as the network usage increased.

Google revealed that their WAN unit costs have been decreasing as their WAN backbone networks have been growing. However, there has been a problem. Their WAN unit costs have not been decreasing fast enough to keep up with the surging increases in WAN bandwidth demand that they have been experiencing [11].

There were multiple reasons for Google's WAN cost problems. As the number of servers and storage devices that were involved in the network increased, the need to have the various boxes talk to one or more other boxes at the same time increased. In order to accomplish this, more sophisticated networking equipment was required. As more and more networking gear was introduced into the network, the need to manually configure each of the separate boxes increased and so more skilled technicians were required. Finally, since multiple vendors were supplying the various pieces of equipment that were being used in the network, trying to automate the configuration of each of the pieces of network equipment using the various non-standard vendor APIs was turning into a significant task.

Table 1 shows the WAN cost components that Google has identified.

Table 1: Google WAN Cost Components

Cost Component Group	Component
Hardware	Routers
	Transport gear
	Fiber
Overprovisioning	Shortest path routing
	Slow convergence time
	Need to maintain SLAs despite network failures
	No differentiation between types of traffic
Operational Expenses / Human Labor Costs	Box-centric vs. fabric-centric views

Google had a great deal of experience with data centers. In the data centers, they had automation tools that allowed them to automate the management of thousands of servers. The tools were so powerful that as more and more servers were added to a data center, the management task did not change significantly. Google was searching for a way to replicate this when it came to managing their network.

What Google wanted to do was to move beyond managing the individual boxes that were used to build their network. Instead, what they wanted to be able to do was to manage all of the boxes at once. Almost as if the network was a type of "fabric" that could be managed as a single entity. The ultimate goal was to be able to manage the network based on the applications that were using it. The goal was to be able to manage application traffic flows instead of network boxes.

Google's desire for a network fabric was thwarted by the equipment that is used to build today's networks. The protocols that are used in networks today to control the forwarding of data packets are all box specific. They have no concept of a network fabric. The network equipment that is being used has been optimized for transferring packets and not for monitoring the network or operations on the network itself. Finally, the network equipment that is in use is designed for the best case – everything is working perfectly. Not enough attention has been paid to the cases where the network is trying to deliver low latency traffic or when there has been a network device failure and a fast failover needs to occur.

9.2 Google's Motivation To Find A Better Networking Solution

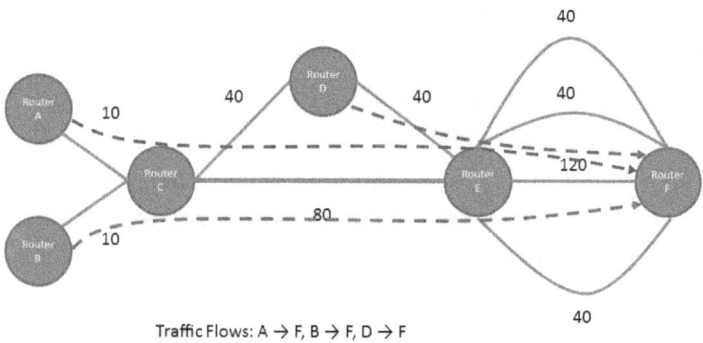

Traffic Flows: A → F, B → F, D → F

Figure 28: Sample Network With 3 Application Traffic Flows

To illustrate the network challenges that Google was facing, Figure 28 shows a sample network with three separate traffic flows. In this network, applications connected to three different routers, A, B, and D all need to communicate with an application connected to router F. Their traffic patterns all take different paths but they share a common network link as they travel from router E to router F.

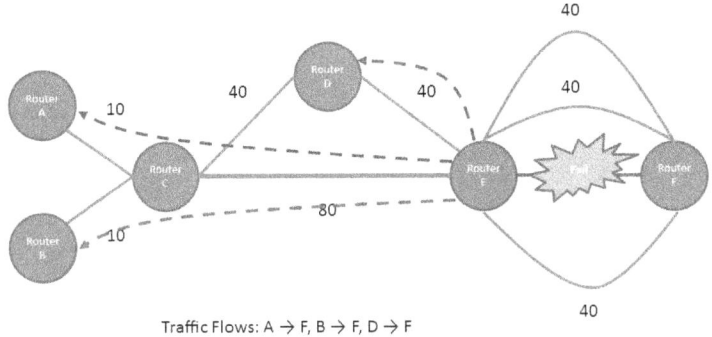

Traffic Flows: A → F, B → F, D → F

Figure 29: Network With Link Failure

Figure 29 shows the same network experiencing a failure on the link that connects router E to router F. This failure will affect the traffic flows for all three of the applications that were using that link to connect to router F. In a modern network, when a failure like this occurs, each of the elements in the network will react to the failure. Router E will immediately send an informational message out to all of the other routers in the network informing them that the link between it and router F has failed. Each of the other routers will then automatously start to take actions to recalculate their flow tables. Each one of the routers who is responsible for sending packets from an application that is attached to it (A, B, and D) will now try to repair the path between it and the destination router, F.

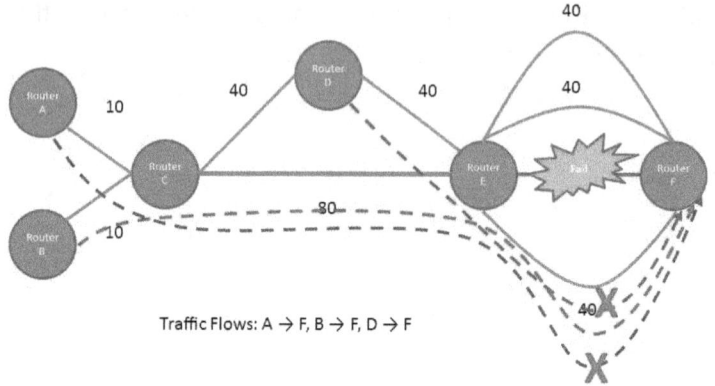

Figure 30: Network With One Traffic Flow Repaired

Figure 30 shows how the network will look if router B is the first router that is able to secure a route to router F that uses the lower link between router E and router F. Note that the route that broke had enough capacity to simultaneously support all three traffic flows. However, now none of the remaining paths from router E to router F can support more than one traffic flow at a time. When router A and router D attempt to use the same path as router B, they will fail.

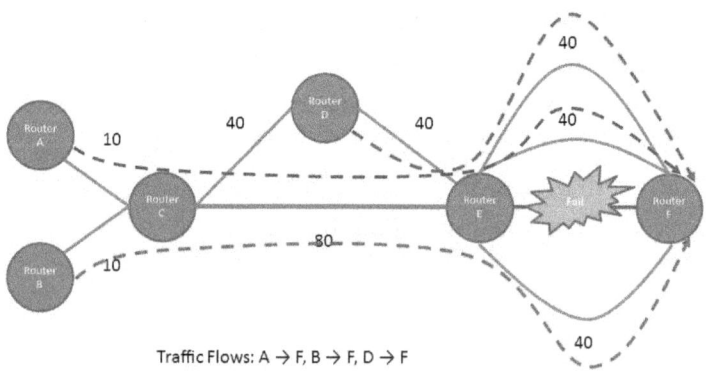

Figure 31: Network after all 3 traffic flows have been rebuilt

Finally, Figure 31shows how the network might look after all three of the application traffic flows had been rebuilt. Google believes that there are two significant drawbacks to the way that this type of recovery from a network failure is done in today's networks. The first drawback is that this process takes time. Routers will attempt to reserve paths, they will fail because other routers have already reserved the path, they will try to reserve another path and they may either fail or succeed. No matter, you cannot definitively say how long it will take a router to rebuild a path to its endpoint when a network failure occurs. The networking term for the time that it takes to rebuild all of the impacted application traffic flows is called "network settling time".

The second drawback to how application traffic flows are rebuilt after a network failure is that the process is not deterministic. Depending on chance, any one of the three routers in our example could have reserved any one of the three remaining routes to router F. There is no way to accurately predict what the network will look like after a link failure. The problem with this approach is that sometimes the network may rebuild a set of links that fully support the needs of the applications that are trying to communicate and sometimes it may not be able to do so. Each of the new links will have various characteristics that may or may not meet the needs of the applications that are using them. It is possible that after a network failure, the network will not be able to create a new connection solution that meets the needs of all of the applications.

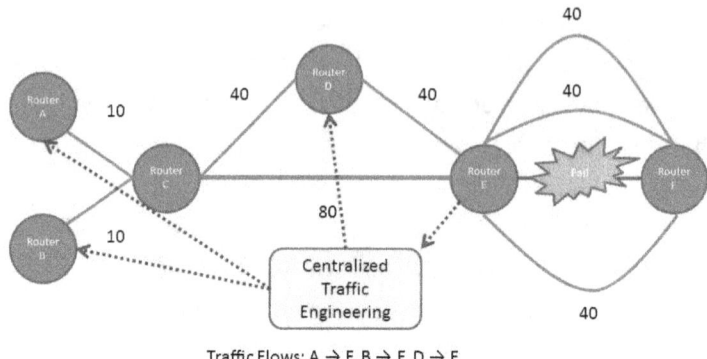

Traffic Flows: A → F, B → F, D → F

Figure 32: Using Centralized Traffic Engineering To Rebuild The Network

What the engineers at Google realized was that there was a better way to manage the link rebuilding process. Figure 32 shows the same network; however, this time it now contains an additional component: a centralized traffic engineering component. In this new scenario, the centralized traffic engineering component is notified by router E when the network link between router E and router F goes down. The centralized traffic engineering component is able to "see" the entire network. Using this information, it can now compute a new set of application traffic flows that will allow the needs of each of the applications to be met and then each of the routers in the network can have their flow tables updated by the centralized traffic engineering component.

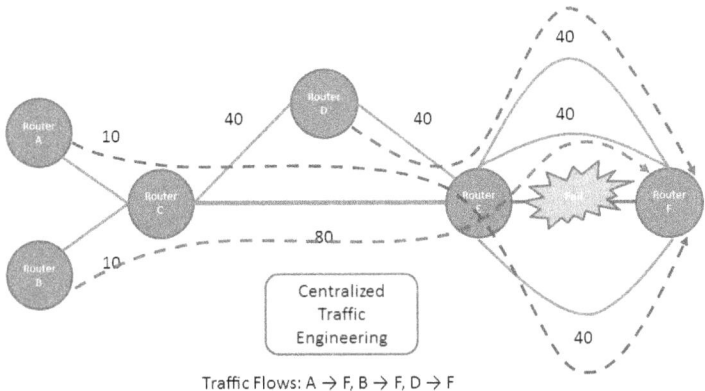

Figure 33: Rebuilt network with centralized traffic engineering

Figure 33 shows how the network could look after having the application traffic flows rebuilt by the centralized traffic engineering component. One important feature of this solution is that it is potentially much faster than the scenario in which each router autonomously built its own new routing solution. Additionally, the network solution is now a deterministic solution – the same new network configuration will be produced each time assuming that the same inputs are provided to the centralized traffic engineering component.

Google was attracted to the idea of creating and using a centralized traffic engineering component for their network because of five key features that it offered:

1. They felt that they would be able to create better network utilization scenarios if the design of the application traffic flows was created using a complete view of the current state of the network.

2. The time that it would take to recover from a network failure was anticipated to decrease because the centralized traffic engineering component would have

all of the required information and would therefore not have to retry solutions. This would greatly decrease the network settling time. The official term for this is "convergence" and it refers to how long it takes the network to compute new routes that will meet the needs of all of the network applications after a network failure has been detected.

3. Google believed that by using this approach they could gain more control over how their limited network resources were being used and this would allow them to decrease the amount of overprovisioning that they would have to do because they would now have deterministic network behavior.

4. The deterministic behavior of the network would now allow Google engineers to create a separate "mirror" network that would allow them to more accurately simulate their production network and test how the various event streams would both act and interact.

5. The computing power of modern routers are by necessity constrained by the cost of the router and the other tasks that they need to be doing at any given time. They can only process new route computations so fast. The centralized traffic engineering component would have no such limitations and was forecasted to be able to compute new network routes up to 50 times faster than a single router could. New configurations could be pre-computed prior to a link failure happening. Google estimated that new routes could be delivered to all routers in the network within one second plus the

propagation delay between the central point and the routers.

9.3 The Importance Of Network Testing

One of the biggest challenges that Google faced with their networks was that they were unable to determine how any given configuration of the network would behave. Given that each router is a complex collection of hardware and software resources, the wide variability of inputs that such a network component could be receiving at any point in time meant that there was no way to predict how the network would react to a given set of inputs.

Google had been trying to create a testbed in which to try out network changes before they were introduced into the network. However, it was acknowledged that in order for any testbed network to provide realistic results, it would need to provide a full scale replica of the production network. This was clearly impossible to do.

However, if a centralized traffic engineering component was used to manage how the network's router traffic flows were calculated, then the current configuration of the traffic engineering component could be duplicated in a test environment. This would allow real production inputs to be used by the Google engineers to both research new ideas and at the same time try out new network configuration plans.

The use of the centralized traffic engineering component has provided the Google test engineers with four advantages that they did not have when they were using the traditional designed network:

1. Testing of network changes can now be done in an isolated fashion. The availability of various logical models lets the Google engineers effectively do "network unit testing" with changes before determining what their impact on the entire network will be.

2. The centralized traffic engineering component has a complete end-to-end view of the network and this allows the Google network engineers to test how their network changes are going to impact the entire network. In this type of simulation, everything involved is real with the exception of the actual network hardware.

3. This type of lab testing allows the Google network engineers to engage in performing built-in network consistency testing. They can cause two network links to fail at the same time in order to determine how the network will react. They can cause a network element to not fail, but start to misbehave and determine what its impact on the network will be. The use of real-world network data allows this consistency testing to be done both in the test lab and in production.

4. The configuration of each device in the network is critical to the correct operation of the network. The ability to test the correct operation of the network has allowed the development of various networking tools that permit the state of each router in the network to be tested. In the lab each router can have its traffic flow table validated after it is updated by the centralized traffic engineering component.

9.4 Simulating The Google WAN

By choosing to implement a centralized traffic engineering component, Google has greatly simplified the task of testing all of their planned network changes. Since all of the router flow tables are calculated in a single location, the Google network engineers are now able to simulate a portion or all of their Wide Area Network (WAN) that interconnects the Google data centers.

Google is now able to treat how they configure their WAN in a similar fashion to how they manage large distributed software projects. In order to ensure that they are able to accurately predict how the production network will react to changes that they plan on making, real production binaries are used in the simulation environment. The actual code that runs the production centralized traffic engineering component is used. Additionally, the actual binary implementation of the OpenFlow interface that will be running on the routers in the network is used also.

In order to be able to simulate a real-world Google network, a very large number of production servers have to be replicated in the testing environment. In order to do this, all of the routers that are being used in the network are simulated in the Google test environment. The real OpenFlow binary code is being used in order to make sure that the simulated switches behave like the real ones will; however, the router's hardware abstraction layer (HAL) is fully simulated. This introduces the limitation that performance of the real world network cannot be accurately measured in the simulation testing laboratory environment.

One of the most powerful features of Google's simulated WAN testing environment is that the Google engineers are able to simulate any arbitrary topology. This includes building an accurate model of the entire Google inter-datacenter backbone network as it currently exists and simulating that configuration. In this simulated network, captured event streams from the real production network can be fed in in order to simulate actual events that happened in the production network.

An important benefit that Google has realized from creating such an accurate simulation environment is that they are now able to test their monitoring software in their simulation environment. Since production code is running in the simulated environment, the monitoring software can be plugged in to the WAN testing environment software and it will produce the same types of results that Google would see in their production environment. The use of the production monitoring software allows the alerting conditions and monitoring software to also be tested in the simulated environment.

9.5 Why Google Is Interested In SDN

Google is a very successful company that has a great deal of funding available to work on almost any project that they so desire. They realize that their enterprise networks are a critical part of their ability to deliver services to their customers and therefore they are willing to invest heavily in any technology that will provide them with a competitive benefit.

Google's engineers have taken the time to investigate the current state of network technologies. They have reached the conclusion that the arrival of Software Defined Networking offers them the greatest return on their investment in the near term. Specifically, they have identified four key benefits that they believe that implementing SDN in their networks can provide to them.

The first benefit is that the implementation of SDN will allow them to separate their network hardware from their network software. No longer will they be required to purchase both products from the same vendor because one will not work without the other.

Instead, Google will now be free to choose their network hardware exclusively based on the hardware features that they feel are necessary. Note that this also means that if there are any hardware features that they do not feel are necessary, they won't have to purchase hardware that has those features. Likewise, network software can now be selected based on the protocol features that Google wants to implement in their network and not based on the hardware that only supports that type of software.

The use of a Centralized Traffic Engineering component will allow Google to logically centralize their network control on a server that can be 25-50 times faster than the processing capability found inside of a standard router. This will provide them with three critical benefits to the performance of their network. When network failures occur, and they will, the network will rebuild its links in a more deterministic time. The rebuilding of links will not suffer from the link congestion issues that require multiple attempts to create a new path between source and destination and therefore will be more efficient. Finally, because the Centralized Traffic Engineer component can have a hot standby, the overall solution will be more fault tolerant.

In today's network, network monitoring functions are provided by the routers themselves. This means that in addition to the packet forwarding and flow table calculations that the router is taking care of, it also has to support management functions. In Google's SDN implementation, this functionality is separated from the router functionality and management, monitoring, and operations functions can be provided independently of the routers themselves.

Finally, Google realizes that the arrival of SDN networking technology brings with it a great deal of promise. We are only at the brink of what can be done when network data is centralized and the network architecture allows it to now be programmed. Google believes that standard software development methodologies can now be applied to creating enterprise networks. This heralds a new era of both networking innovation coupled with enhanced flexibility that promises to change how networks are built and managed in the future.

Google believes that by implementing SDN technology at the heart of their mission critical enterprise networks they will be able to achieve multiple benefits. They expect to have a WAN that is not only able to provide them with higher performance, but at the same time will be more tolerant of network faults and should turn out to be cheaper to both build and maintain.

9.6 Google's Open Flow WAN

When Google was considering how they would implement a SDN network, they started by taking a look at their enterprise networks. Google operates two separate backbone networks as a part of its business.

The first network is referred to as being "I-Scale" ("I" for Internet?) and it is the backbone network that is Internet facing and carries user traffic for services such as Google search, YouTube, Gmail, etc. This network looks very similar to any ISPs network and has been constructed using a standard array of commercially available networking gear from the leading network equipment vendors.

This network, in the words of Google, must be "bulletproof" because it is providing services to end users. It has demanding availability and loss sensitivity requirements that it must meet.

The other backbone network that Google operates is called the "G-Scale" ("G" for Google?) backbone network. This network is used to interconnect Google's data centers. Although it too has demanding network requirements, because it is used for tasks like server backup and search index transport, there is more flexibility in how it performs. The traffic on this network is much more bursty than the traffic carried by the I-Scale network [11]. The network consists of 12 worldwide data centers that are interconnected over 10G links [14]. This was the network that Google selected to perform its SDN experiment on.

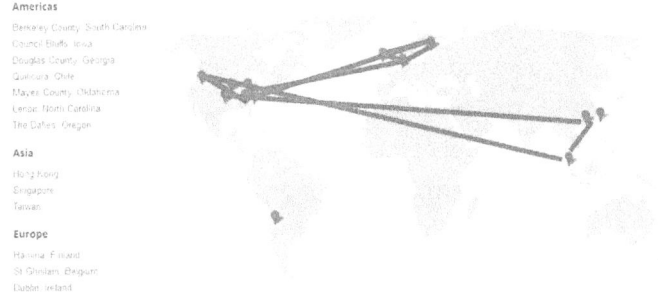

Figure 34: Google Data Center Locations

Figure 34 shows the locations of Google's current set of data centers.

9.6.1 Google's G-Scale Network Hardware

The hardware that Google uses for the routers that have been used to build the G-Scale backbone network are custom designed configurations because when the networking project was started in 2009 / 2010 it was not possible to purchase a router that supported the OpenFlow protocol. The most important thing to realize about these routers is that despite the fact that they are being used by a very financially successful company to build a critical enterprise network, there is nothing unique about them – no customer ASIC chips have been used.

The G-Scale routers have been built using off the shelf silicon. Each of the routers is able to provide the network with 128 ports of non-blocking 10GE connectivity. Since the routers are being used as a part of a SDN network, each router supports an OpenFlow networking protocol agent. What makes these routers somewhat unique is that outside of the software needed to boot the box and the OpenFlow agent, they run very little other software.

The routers do not even support a command line interface (CLI). Everything, including the CLI, runs on the central control server which is a multi-core sever with a great deal of memory. Additionally, the central controller contains open source software stacks that permit the support of the Quagga network routing software suite which allows the router to support both the Border Gateway Protocol (BGP) and the Intermediate System-to-Intermediate System (ISIS) protocol. When BGP runs between two peers in the same autonomous system, it is referred to as the Internal BGP (iBGP or Interior Border Gateway

Protocol). The ISIS / IBGP protocols are used for internal connectivity. However, because Google could specify what functionality the routers have, the central controller doesn't support unneeded networking protocols such as AppleTalk and MPLS (which was supported on Google's legacy G-Scale network).

In order to provide the network bandwidth that the G-Scale backbone needs, each Google data center contains multiple chassis. This allows fault tolerance to be built into the network by having chassis provide backup for other chassis. Additionally, support for multiple chassis per site allows the G-Scale backbone to scale to multiple Terabytes of networking capacity.

Figure 35: Google's G-Scale Backbone Network WAN Deployment

Figure 35 shows how Google's data centers were interconnected using the G-Scale backbone network. In each data center there are multiple Google OpenFlow switches which are cross connected with each other. A set of long-haul optical transmission links then connect routers in one data center to the routers in the other data center. The routers are custom hardware configurations that are all running the open source Linux operating system.

9.6.2 How Google Rolled Out Their SDN Network

Google was faced with a difficult, but not unique, challenge when they decided to introduce the SDN network technology into their G-Scale backbone network. The functionality that this network provides to Google is not customer facing; however, it is critical to the 24x7 operation of the company. The network could not be completely taken down in order to retrofit it with the SDN technology.

In the spring of 2010 Google started Phase 1 of their rollout of their SDN solution. The first step was to introduce the new OpenFlow controlled switches into the network in parallel with their existing network hardware in three of their data centers. Their plan was to make the new switches to look like the existing routers that had already been deployed in the network.

What this meant was that there was no change in operations from the perspective of the G-Scale's existing non-OpenFlow switches. The BGP / ISIS / OSPF protocols were used to interface to the centralized OpenFlow controller in order to program the state of each of the controlled OpenFlow switches.

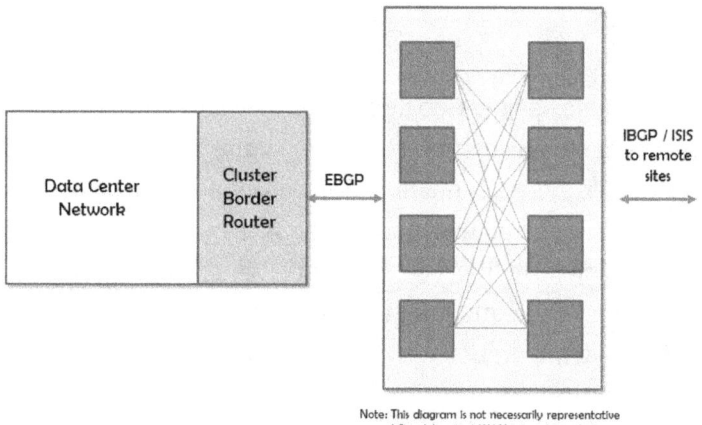

Note: This diagram is not necessarily representative of Google's actual WAN network topology

Figure 36: Mixed SDN Deployment

Figure 36 shows what a typical data center network component would have looked like prior to the introduction of the SDN network components. In this depiction of the data center network, at the edge of the network there exists a cluster border router. This border router communicates with the Google core network using the EBGP protocol. This is the protocol that is used to provide connectivity between Google data centers. The cluster border router is peering with a group of switches at the edge and these switches are then connecting across the WAN to switches at other data center sites. The IBGP / ISIS protocols are then used for internal communication between different data centers.

Figure 37 shows the same network as before; however, now a set of SDN network control applications have been added. These applications were added on a separate server in order to move time-critical network protocol calculations off of the processors that are embedded in the network routers and provide them with their own high speed computation platform. This set of control applications includes the follow applications:

- Quagga: A suite of network routing applications that provide an implementations of Open Shortest Path First (OSPF), Routing Information Protocol (RIP), Border Gateway Protocol (BGP) and IS-IS for Unix-like platforms.

- OFC: the Open Flow Controller (OFC) manages the OpenFlow protocol. OpenFlow is a communications protocol that is used to give access to the forwarding plane of a network switch or router over the network.

- Glue: allows the open flow controller (OFC) to talk with the Quagga application

- Paxos: a family of protocols for solving consensus in a network of unreliable processors. Google uses Paxos for leader election. The Paxos family of protocols includes a spectrum of trade-offs between the number of processors, number of message delays before learning the agreed value, the activity level of individual participants, number of messages sent, and types of failures. Paxos is usually used where durability is required (for example, to replicate a file or a database), in which the amount of durable state could be large. Based on their real-world experience, Google recommend that an odd number of Paxos instances be used for leader election.

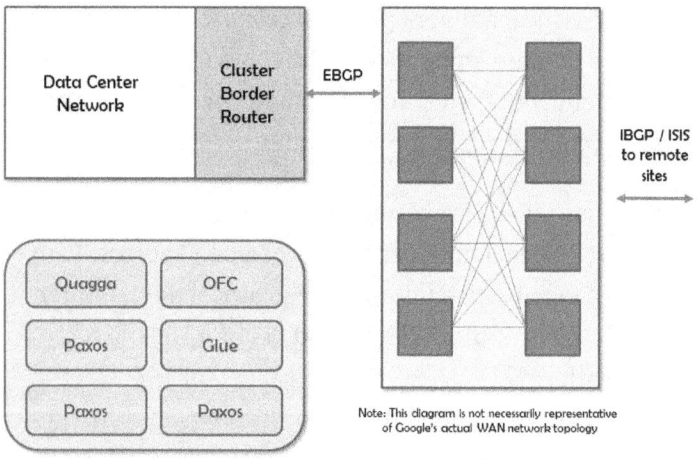

Figure 37: Mixed SDN Deployment With Control Applications

The next step in the process is shown in Figure 38 in which the protocols that the new application server will use to communicate with the rest of the network are added. This allows OpenFlow agents to start to be run on a subset of the switches that are controlling the network. The OpenFlow controller can be connected to the OpenFlow agents using a variety of methods including upgrading half the network at one time, doing every other switch, or just doing a pilot set of switches initially.

Once this change was made, Google had effectively divided their backbone network into two parts. The first part was the traditional backbone network. In this network, the network uses the EBGP protocol to talk to the cluster border router. ISIS can be used to communicate with other WAN sites. There is no OpenFlow protocol being used in this configuration.

The other half of the network that consists of the new OpenFlow nodes is now speaking EBGP through Quagga on the high speed server. Additionally, IBGP / ISIS is being spoken through Quagga on the high speed server.

Forwarding table entries on the individual switches in the network are being coordinated through the OpenFlow agents. Quagga is able to do this because of the information that it gets from the OFC application via the Glue application that is used to connect Quagga and the OFC.

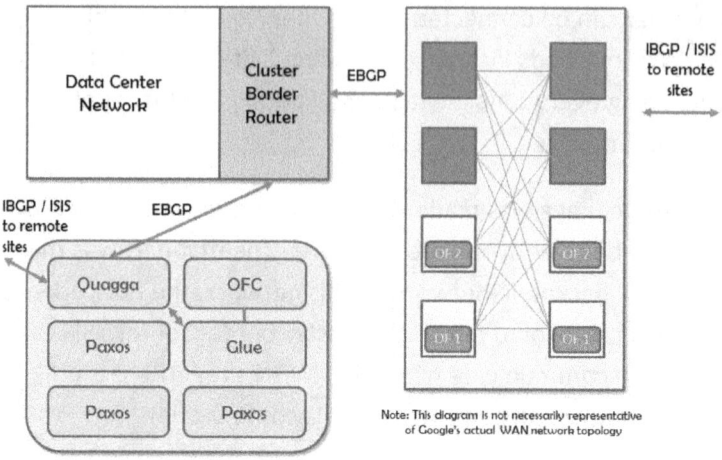

Figure 38: Adding Protocols To Allow Server To Communicate With Network

The process that the Google engineers used to deploy the new OpenFlow switches was performed on a site-by-site basis. This meant that the entire backbone network didn't have to be upgraded all at once and for that matter, an individual site within the network didn't have to be fully upgraded all at once. The key to this upgrade technique was that the applications that were running in Google's date centers would see no differences between the OpenFlow routers and the non-OpenFlow routers.

The first step was to pre-deploy the gear at a given site. Next, 50% of the site's bandwidth would then be taken down. The engineers would then perform the needed upgrades. Next the bandwidth would be brought up using the new OpenFlow switches. As a final step, testing of the new functionality would then be performed. Some of the network's optical data center interconnections were then moved over and made available to the new OpenFlow routers. As the new SDN functionality was introduced, the sites that had been upgraded to SDN provided full interoperability with the legacy router sites. This process was then repeated at a total of three of the G-Scale network's sites.

In order to test the new network that had been created, the Google engineers very carefully selected applications that they would allow to make use of the OpenFlow switches and the associated network bandwidth. Low risk applications, such as a copy service were permitted to opt in and use the new network.

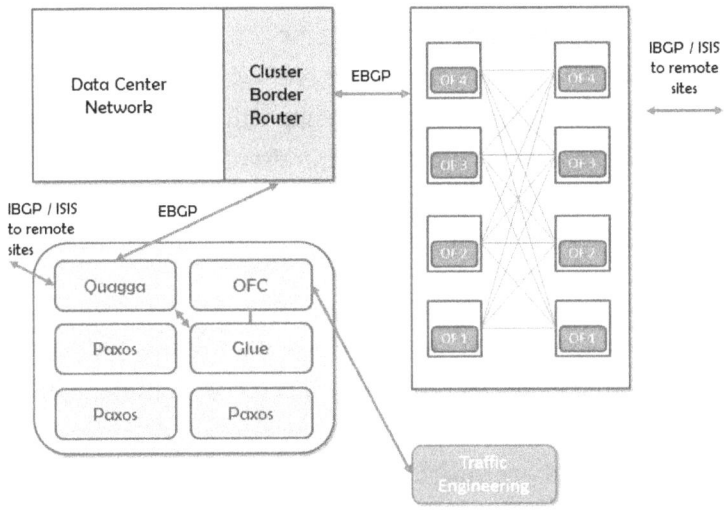

Figure 39: Adding Additional Functionality TO SDN Network

Once the entire backbone network had been upgraded to use the new SDN technology, the functionality of the legacy network had simply been duplicated. The next step was to add additional functionality (as shown in Figure 39) . The first step in this process was to add a traffic engineering server that interfaced to the OFC module of each of the SDN routers. This traffic engineering server was then able to have a global view of the communication patterns that were being used in the backbone network.

Phase 2 of the deployment of the OpenFlow technology in the G-Scale network took roughly another year. The ramp up lasted until mid-2011. During this time the Google engineers proceeded to activate a simple SDN network with no traffic engineering (TE). The network was extended to all of the G-Scale's sites; however, the SDN network was still being operated in parallel with the traditional G-Scale network.

In order to continue to test the performance of the new network, more and more internal traffic continued to be moved over to the OpenFlow switches. This real-world network provided the Google engineers with opportunities to test their ability to transparently introduce centralized controller software updates without impacting network traffic. Additionally, failover of the centralized controller was tested.

Phase 3 of the project started in early 2012. As part of this phase, the SDN network became part of the full production G-Scale network at one site. Once this was successfully done, the rollout continued to additional sites and now all of Google's data center backbone traffic plus some of the Internet facing traffic that was found to be suitable for this network is carried by the new OpenFlow network. The old network that had been built with legacy switches has been turned off.

The centralized controller has been upgraded via a major software release. Now the centralized controller supports traffic engineering functionality. This means that when it is establishing network paths, it can take into account the bandwidth needs of the various applications involved and will ensure that each application will get the bandwidth through the network that it needs. Routing is optimized based on seven application level priorities. This allows the centralized controller to provide globally optimized placement of network flows.

As an example of the new types of functionality that Google has been able to implement because of their use of the SDN technology in the G-Scale network, they can now support bandwidth reservations. The Google application that schedules the creation and transport of copies of large data sets is able to interact with the central OpenFlow controller. This interface allows the copy scheduler to work with the OpenFlow controller in order to implement deadline scheduling for large data copies.

9.6.3 Bandwidth Brokering and Traffic Engineering In Google's Backbone Network

The implementation of the SDN backbone network technology changed the way that Google was able to manage the bandwidth used in their G-Scale backbone network. Specifically, the new technology now allowed them to both broker and engineer the way that bandwidth was being used on a global basis.

	App1	App2	App3	App4	App5	App6	App7	App8	App9	App10
App 1	--	27	--	--	--	65	--	71	24	18
App 2	78	--	22	71	35	71	53	66	12	--
App 3	--	--	--	82	50	--	--	--	--	8
App 4	24	92	16	--	83	27	26	56	--	--
App 5	5	51	60	84	--	9	12	20	42	79
App 6	93	13	17	--	78	--	20	90	3	--
App 7	--	57	--	51	--	87	--	69	--	--
App 8	39	21	80	63	22	34	43	--	66	10
App 9	38	30	22	56	8	79	51	26	--	61
App 10	--	--	--	--	47	--	--	--	--	--

Figure 40: Network Bandwidth Requests For Communication Between Applications (in Mbps)

In a typical Google data center, at any point in time there are a large number of applications that are running at the same time. Each one of these applications would like to have a portion of the Google G-Scale backbone network's bandwidth allocated to it so that it can use it to communicate with some other application. Figure 40 shows an example of the types of bandwidth requests that could exist between a subset of the applications running in the network. Ultimately, there is more demand for network bandwidth than there is bandwidth available to use. Additionally, not all of these applications have the same level of importance. Some applications, such as backups, need to be done but don't necessarily need to be done right now.

This means that the limited amount of G-Scale network bandwidth needs to be brokered among the various applications that are requesting it. The high priority applications will, of course, receive more bandwidth to use and the lower priority applications will receive either no bandwidth or less bandwidth.

The network bandwidth will be allocated among the various applications using Google business rules by the Traffic Engineering & Bandwidth Allocation Sever that is shown in Figure 41. The Bandwidth Requests Collection & Enforcement portion of this server is responsible for collecting the G-Scale network bandwidth requests from all of the various applications that want to use the network. This module has the ability to implement rate limiters on applications – if they have exceeded the amount of bandwidth that the Google business rules had allocated to them, then the amount of bandwidth that they will receive in the future could be reduced.

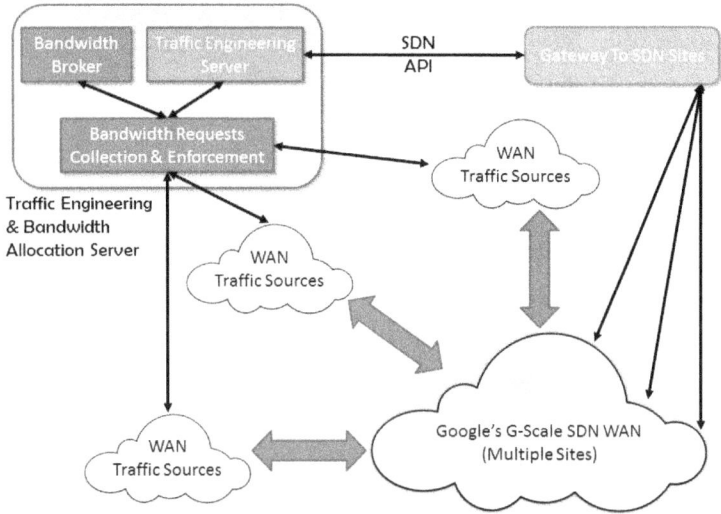

Figure 41: Google's SDN Network High-Level Architecture

All of the bandwidth requests are then forwarded to the Traffic Engineering Server module. It is the job of this component to create a solution for how application bandwidth should be allocated in the network. The bandwidth allocation results of the traffic engineering server will then be communicated via a SDN API to a gateway to the SDN network sites. This information will then be used to program the forwarding tables in each of the SDN routers in the network.

Figure 41 shows the architecture of the bandwidth broker component that is being used in the Google network. This is then used to communicate global demand to the network's traffic engineering server application.

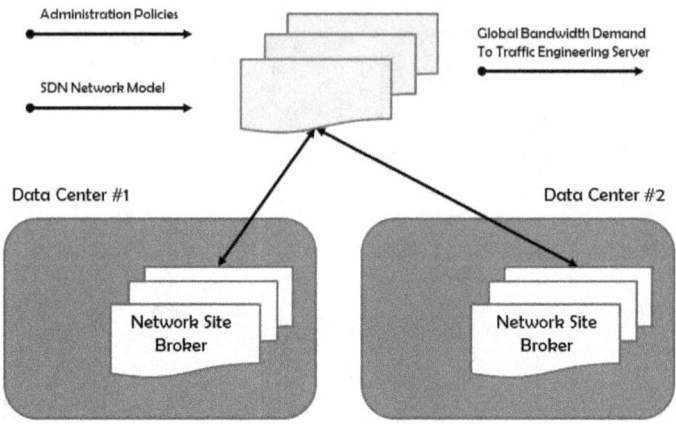

Figure 42: Architecture of the Bandwidth Broker

Figure 42 shows the service architecture of the traffic engineering server used in the G-Scale network. The network's global bandwidth broker module starts the traffic engineering process by passing it a "bandwidth demand matrix" which consists of source and destination applications along with their requests for bandwidth. Note that this is similar to what was shown in Figure 40.

In a legacy network, the path that would be constructed to connect two applications that wanted to talk with each other would be based on the shortest path first algorithm. In Google's SDN network, the Path Allocation Algorithm is able to determine based on a non-shortest path first basis what the optimum route through the network would be in order to connect two applications that want to talk to each other.

This route will be calculated by the traffic engineering module based on its real-time knowledge of the network's topology. Any known network failures will have already flowed up through the network and will already be incorporated into the path allocation algorithm.

The end result of the traffic engineering module will be a set of path assignments. This will indicate which source should talk to which destination using which path. This will be the information that is then sent to the gateway by the SDN sites that was shown in Figure 41.

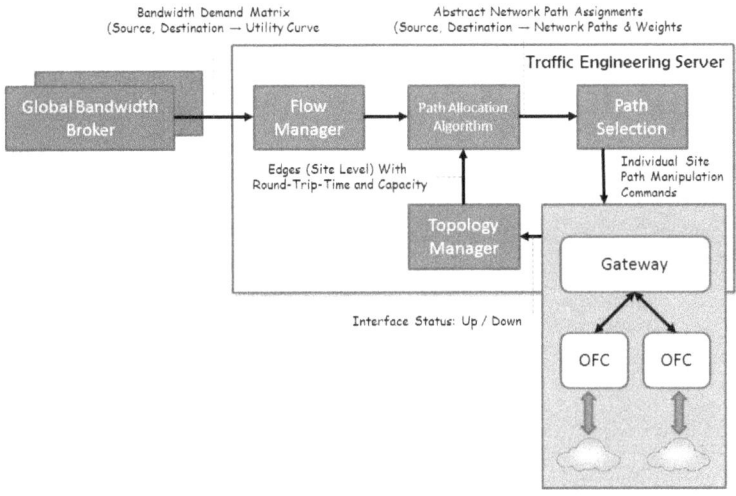

Figure 43: Traffic Engineering Server's Service Architecture

Figure 43 shows what is happening in a specific Google data center. Each one of the switches is executing an OpenFlow Agent (OFA) application. In this data center there is an OpenFlow Controller (OFC) which is charged with managing multiple SDN switches.

The controller is receiving inputs from two different types of sources. The first of these sources types has to do with routing: BGP and ISIS protocol information. These sources will be providing the controller with the default shortest path information. This is basically the routing information that it would have if there was no traffic engineering functionality in the network. The second source is a tunneling application that will be providing traffic engineering operations information to the controller.

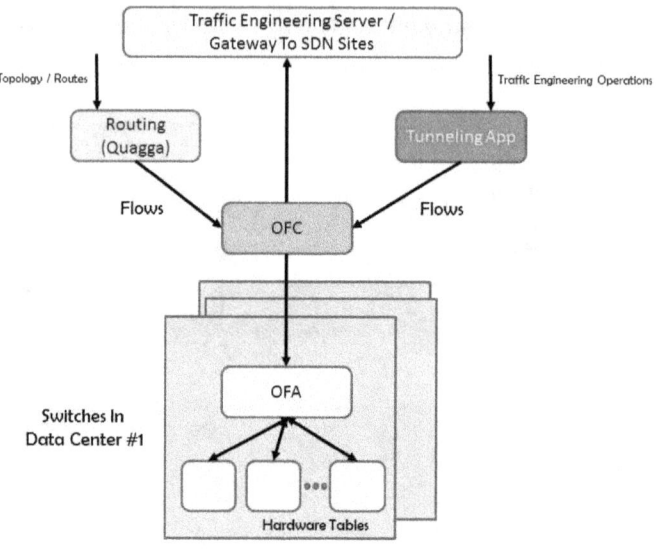

Figure 44: Architecture Of The SDN Controller

In Figure 44 a sample network site with three SDN routers. An application that is connected to router #1 wants to connect to an application that is connected to router #3. Using the information that has been provided via the routing information source, router #1 could be programed by the site's controller to create a direct path from router #1 to router #3 – this would correspond with the shortest path first solution. The applications that would have their information sent over this path would be the ones that were either high priority or which had connections that were latency sensitive.

The information that has been provided via the traffic engineering system via the tunneling application can then be used to set up other network routes from router #1 to router #3 that pass through other routers in the network. Note that these routes will not be shortest path first routes. The applications whose information that would be sent over this path may not be as high priority or may not be as latency sensitive.

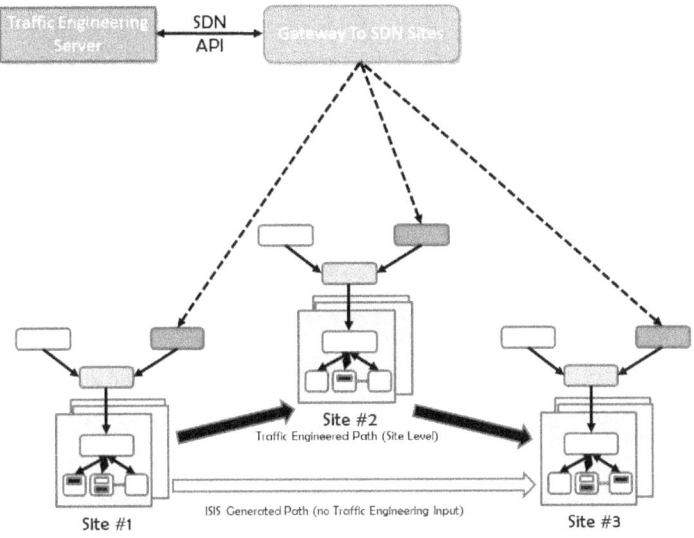

Figure 45: Use Of The SDN Controller In Google's Backbone Network

9.6.4 Results Of Google's Deployment Of SDN

One of the important discoveries that Google made as a result of introducing the SDN functionality into their network was that they were able to achieve a much faster introduction of new network functionality or bug fixes. Their production-grade traffic engineering functionality was added to the SDN network in just two months.

There are two reasons for this increase in speed. The first is that because they had fewer devices that needed to be updated. Instead of hundreds of routers that are currently processing network data and have to be updated on the fly, now only a few control servers have to be updated.

Updating applications is a key part of how Google runs its main business, so they have many existing tools that they can use to perform this function. Hitless software upgrades and new feature additions to the G-Scale OpenFlow network are able to be performed. These changes do not cause any packet loss or any network capacity degradation while they are being performed. One of the reasons for this is because most of the feature releases do not impact the actual network switches themselves.

Additionally, the Google simulation environment allows the entire network to be simulated. This functionality allows the Google engineers to do much more extensive testing prior to rolling out the new functionality. One discovery that they made is that if a solution for placing all of the required network flows exists, the centralized network controller will find it. In a traditional network, a new network flow solution might theoretically exist, but because of how the protocols worked independently the correct solution would never be found.

Google has stated that their ultimate goal is what they refer to as "push on green". This means that when they have a software build that passes all of its submit tests, they should be able to instantly push it out into the production environment without any delay.

Security is an important part of operating any network. Google believes that the SDN network is more secure than a traditional network. The reason for this is because in an SDN network, it is only the central controller that needs to be secured. This controller can be isolated from the rest of the network and can be restricted to only talking with the switches in the network. These communication sessions can be secured though the use of certificates. In a traditional network, if any one of the routers was compromised because someone was able to log into a switch, then the entire network would be at risk because additional routes could be inserted into its flow table. This is not a problem in a SDN network in part because there no one is permitted to log into the network switches.

Google reports that as a result of implementing the SDN network they are already seeing a higher utilization of the G-Scale network. They have seen network utilizations of the links that connect their data centers together that were close to 100% for extended periods of time (days). This is to be compared to traditional legacy WAN backbone networks that normally experience link utilizations of 30% - 40% [11].

The reason that this high degree of network utilization is occurring is because the new SDN technology allows for flexible management of end-to-end paths for maintenance. Google believes that you could "tune" a traditional network to provide the same type of deterministic network planning results that they are able to achieve with the SDN network. However, the amount of time that this would require makes it impractical to do in the real world.

An additional benefit of the use of the SDN technology is that more application data can be exchanged via the network and less operations data has to be exchanged. What Google has discovered is that in their SDN backbone network they have had a significant decrease in the amount of protocol related traffic that they need to transport. When the volume of tunneled traffic engineering changes is compared to the traditional amount of ISIS protocol traffic that the network has to carry, Google was able to report a 6x decrease in the amount of traffic that was transported by the network [11]. They attribute this savings to the fact that they no longer need to operate at the link level of network granularity.

The stability of the network was very high. The service level agreement (SLA) for the internal backbone network is being met by the new SDN network. At the completion of Phase 3, Google had experienced one network outage from a software bug and one network outage that had been triggered by the push into production of a "bad" software configuration.

Google is not done with their SDN network. In fact, they admit that they are really just getting started with it. As they move forward they see several different immediate opportunities that the new SDN technology is going to be able enable for them.

Google believes that with the SDN technology it will be much easier to understand what is going on in their network. They will be provided with a unified view of the network fabric. If they want to know what would happen under a given set of circumstances, then all they have to go is to go to their simulated network testing environment and they can easily get an answer to their question. This insight is aided by the enhanced predictability that the network now provides. A Google network engineer can now take a look at the flow tables and understand how a given flow was created.

The traffic engineering functionality that Google has been able to implement in the SDN network has yielded a number of real-world benefits. The first of these is that they now have a higher awareness of the quality-of-service (QoS) that the network is providing along with enhanced predictability of failures.

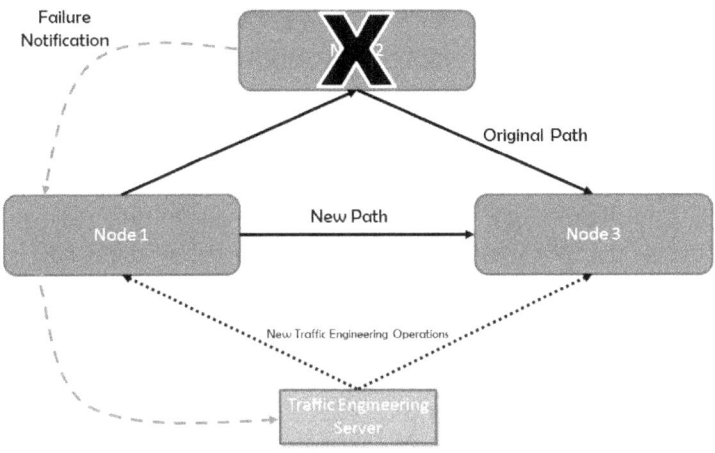

Figure 46: WAN Convergence Under Failure Conditions

Figure 45 shows a sample 3 node network that is experiencing a network failure of node #2. In a legacy network that did not have the traffic engineering functionality that the SDN can support, the link failure could take up to 9 seconds to resolve. Both the detection of the failure and the convergence to a new network configuration solution would both take longer than in a SDN network.

Google believes that the delay within the traffic engineering server in a SDN network would be less than the ISIS protocol timers that would be used to detect and the communicate the failure. There are work-arounds to these delays, such as "fast failover", but the network path solutions that they will create are not guaranteed to be accurate or optimal. It is Google's experience that in a SDN network this type of network failure can be resolved with an optimal new network configuration within 1 second [11].

The new SDN network is providing enhanced packet latency, packet loss, bandwidth utilization, and deadline sensitivity measurements. Additionally, no longer are all network applications treated the same. The SDN network allows Google to differentiate between applications and their unique network needs.

The SDN network is able to provide Google with improved routing between switches for applications. This is able to be provided because the SDN network's central controller has a priori knowledge of the network's topology and the L1 / l3 connectivity that exists in the network.

Finally, Google has been able to create software tools that provide them with improved network monitoring and alerts. The use of a centralized controller allows many of the monitoring tools to be connected to this high speed server which is able to provide network routing data in real time.

9.6.5 Challenges Of Implementing A SDN Network

Although Google was able to successfully implement and operate a SDN network, they did discover a number of challenges with this new technology as a part of their project. One of the most significant issues that they encountered was that the OpenFlow protocol is still in its infancy. Not all of the functionality that they would have liked to have had has yet been incorporated into the OpenFlow protocol specification. However, Google reports that what is there is good enough to successfully build and operate a network.

When they implemented their SDN network, Google chose to not control it from a single NOC. Instead, network control is replicated and distributed in order to boost the network's fault tolerance. Google views having replicated distributed control is a fundamental design requirement. [14]

The importance of the centralized controller cannot be overstated. Considering the critical functionality that is provided by this network device, a backup version must always be available. In practical terms this means that control plane functionality has to be implemented that will allow the various centralized controller boxes to elect who is currently the master controller. Correctly implementing this logic is difficult and complex to do. It can be done, but it requires both effort and time.

Another important question that has to be answered is where software functionality should be located: on the switch or on the centralized controller. Google's plan is to move as much functionality to the centralized controller as possible, but some must still be left on the switch. What to leave and how to configure what has been left are inexact sciences.

In very large networks, there will be many different traffic flows. The occurrence of a major configuration change in the network means that most if not all of those traffic flows will now have to be recomputed. Initially Google ran into performance problems when they tried to recalculate the traffic flows quickly enough to not impact network performance. These problems were eventually solved.

Google attributes their ability to introduce the SDN network so quickly to their new found ability to write and introduce network software so quickly when it's being used in a standard environment on a powerful centralized server. This is in sharp contrast to trying to write software for the router environment which is very limited in terms of functionality and processing power.

9.6.6 Google's Conclusions After Having Implemented A SDN Network

Having completed the conversion of their G-Scale backbone network to use SDN technology and retiring the legacy networking gear that had been used to provide the network, Google has been able to reach some conclusions about the use of SDN technology.

The first of their conclusions is that the OpenFlow protocol, although not complete, is ready for use in real-world networks. Google feels that more functionality is needed; however, this should not hold network designers back from creating new networks based on the OpenFlow protocol.

The SDN network architecture with its use of a centralized network controller is an idea that is also ready for real-world use. Google feels that networks that are implemented using this new technology will permit the rapid deployment of rich feature sets. Network operators will also benefit from SDN's simplified network management operations.

The G-Scale network is a relatively simple network. It only connects a handful of Google data centers. It is Google's belief that if SDN technology is applied to a more complex real-world network, then the benefits to the network operator would be even greater.

In the short term, Google believes that more standardization is needed. Specifically, they are looking for ways to reduce the amount of time it takes to program a large network. Additionally, Google would like to be able to better determine how to configure the hardware that remains needed. [13]

Google has fully committed to the use of SDN within its enterprise networks. Their mission critical inter-data center G-Scale backbone network successfully runs on the OpenFlow protocol. This network is the largest production network in terms of traffic carried operated by Google. The implementation of the SDN technology has resulted in improved manageability and they are anticipating that over time it will also provide improved cost savings.

Google measures the value of implementing the SDN technology in several different ways. One is the improvements in the utilization of the network bandwidth. This is a significant cost savings to Google and will help to amortize the cost of the development of the new OpenFlow switches and the centralized controller software.

This increase in utilization is being provided with potentially better guarantees of service even under network failure conditions due to the rapid recovery of the SDN network. Google also believes that they will be able to operate their G-Scale backbone network at the same level of service as they did the legacy network with much less effort (and cost) on their part.

These savings should come about because of the reduction in configuration and monitoring that the SDN network will require. The ability to have the network automatically react to events by itself with little or no human interaction required should further reduce the costs of operating the SDN network. Google anticipates that their savings will be measured in terms of unit cost / Mbps / month of bandwidth delivered for a given SLA.

10 OpenFlow Topics

The OpenFlow protocol structure communicates between the control and data planes of supported network devices. OpenFlow™ has been designed to provide an external application with access to the forwarding plane of a network switch (or router). Access to this part of the router can be gained over the network which allows the controlling program to not have to be collocated with the network switch.

Traditional networking protocols have tended to be defined in isolation with each solving a specific problem and without the benefit of any fundamental abstractions. The result of this isolation has been the creation of one of the primary limitations of today's networks: complexity.

As an example of this complexity, in order to move a device from one location on the network to another location on the network the networking professionals must touch multiple switches, routers, firewalls, Web authentication portals, etc. and update ACLs, VLANs, quality of services (QoS), and other protocol-based mechanisms [19] using network management tools that operate at the device and link levels.

Additionally, when these types of changes are being made, the network topology, vendor switch model, and software version all have to be taken into account. The end result of this network complexity is that once a network is built, it basically stays as it is so that nothing becomes broken.

The OpenFlow protocol has been created to solve the problems that legacy networking protocols have created. In an SDN architecture, OpenFlow is the first standard communications interface defined between the control and forwarding layers.

OpenFlow allows direct access to and manipulation of the forwarding plane of network devices such as switches and routers, both physical and virtual (hypervisor-based). Currently, no other standard protocol does what OpenFlow does, and it has been determined that a protocol like OpenFlow is needed to move network control out of the networking switches to logically centralized control software.

When the OpenFlow protocol is implemented, it is implemented on both sides of the interface between the network infrastructure devices and the SDN control software. In order to identify network traffic, OpenFlow uses the concept of flows based on pre-defined match rules that can be statically or dynamically programmed by the SDN control software.

OpenFlow allows network professionals to define how traffic should flow through network devices based on parameters such as usage patterns, applications, and cloud resources. OpenFlow allows the network to be programmed on a per-flow basis and so this means that an OpenFlow-based SDN architecture can provide extremely granular control, enabling the SDN network to respond to real-time changes at the application, user, and session levels. In today's legacy networks, IP-based routing does not provide this level of control, as all flows between two endpoints must follow the same path through the network, regardless of their different requirements.

The OpenFlow protocol has been created to enable software-defined networks and as of today is the only standardized SDN protocol that permits direct manipulation of the forwarding plane of network devices. OpenFlow was initially applied to Ethernet-based networks; however, OpenFlow switching can extend to a much broader set of use cases. OpenFlow-based

SDNs can be deployed on existing networks, both physical and virtual. Network devices can simultaneously support OpenFlow-based forwarding as well as traditional forwarding. This means that it is easy for enterprises and carriers to progressively introduce OpenFlow-based SDN technologies, even in multi-vendor network environments.

10.1 An Overview Of The OpenFlow Switch Specification

The Open Networking Foundation (ONF) is a user-led organization dedicated to promotion and adoption of software-defined networking (SDN). The ONF manages the OpenFlow standard. Version 1.1 of the OpenFlow protocol was released on February 28, 2011. The next version of the standard (ver 1.2) was published in February 2012. The current version of OpenFlow as of the writing of this book is ver 1.4.

The ONF OpenFlow protocol defines what the requirements of an OpenFlow switch (shown in Figure 46) are. An OpenFlow switch consists of three main components: an OpenFlow Channel, a Group Table, and one or more Flow Tables. The flow tables and the group table are responsible for performing packet lookups and packet forwarding. The OpenFlow channel is used to communicate with an external controller. The external controller uses the OpenFlow protocol to manage one or more OpenFlow switches.

What happens in an OpenFlow switch at a high level is that Ethernet frames ("packets") arrive and their headers are compared to data that has been stored in a flow table within the switch. Every OpenFlow switch must contain one flow table and can contain multiple flow tables. If a match between the contents of the packet's header fields and the flow table entry is made, then a set of instructions is then executed. Each one of the switch's flow table entries contains three pieces of information: the data that is used to match the fields in the received packets ("match fields"), counters that keep track of the number of matches that have been made, and the instructions that are to be executed if a match is successfully made.

The matching of the fields in a packet starts with the first table (numbered "0") and continues on until a match is made or a table-miss event is declared to have happened. Exactly what happens when a table-miss event occurs is dependent on how the switch is configured, but options include dropping the packet, forwarding it to the controller for further processing over the OpenFlow channel, or continuing on to the next flow table in order to continue the search for a match to the packet.

As the network changes, the OpenFlow switch's flow tables have to be updated. Updating these tables is the responsibility of the external controller that the switch is connected to.

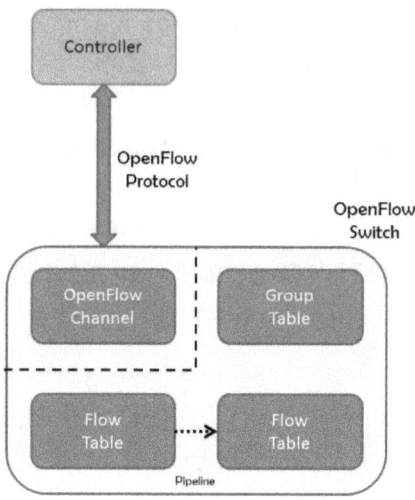

Figure 47: OpenFlow switch main components [20]

If a packet reaches the end of a flow table and it still has not been matched, then if the table-miss instructions modify the packet processing pipeline the packet may be allowed to be sent to the next flow table in the pipeline. Anytime the instructions that have been retrieved from a flow table entry because of a packet match or a table-miss event do not specify a next flow table, then processing of the packet will come to a halt. The packet will generally then be modified and forwarded on to the next switch.

The OpenFlow switch specification is just that – a specification. This means that the designers of switches have the ability to implement the OpenFlow functionality in any fashion as long as it conforms to the standard. The functionality of the switch can be split between hardware and software in any fashion that the designer chooses to do so.

10.1.1 OpenFlow Ports

Ports are a critical part of the OpenFlow protocol because they specify where a packet comes from and ultimately where it will be going. When two OpenFlow switches are connected, they are connected via ports.

An OpenFlow switch contains three different types of ports: physical, logical and reserved. Each of these port types behaves differently. Any one of these port types can be used as both an input and an output port for a packet.

Physical ports are exactly what they sound like – ports that correspond to a physical interface port on the OpenFlow switch hardware. Logical ports are not related to a physical port on the switch. However, logical ports can be made to map to a physical port on the switch. When packets are being processed by the OpenFlow switch, both physical and logical ports are treated exactly the same way.

Reserved ports are special ports that are used to cause a specific action to occur. This action is triggered by sending a packet to a reserved port. An OpenFlow switch is required to support 5 types of reserved ports. There are 3 additional types of optional reserved ports that can be supported by the switch.

10.1.2 OpenFlow Flow Tables

At the heart of any OpenFlow switch is its flow tables. These tables are used to determine what, if any, action is to be taken based on the receiving of a given packet. The flow tables are an important part of the OpenFlow switch's packet processing pipeline.

10.1.2.1 OpenFlow Packet Processing Pipeline

Figure 47 shows how the set of flow tables in an OpenFlow switch work together to process a received packet. Every switch must have at least one flow table. This table is assigned the number "0". Additional flow tables can exist and each of them has an identifying number. A received packet will first have the contents of its header bits matched to the flow entries in table 0. In the case that a match is made, then the instructions that are part of that flow entry will be executed. One of the instructions that is executed may direct the OpenFlow switch to send this packet to another table (that table must have a larger table number than the table that is currently processing the packet) in order to continue to attempt to make more matches between flow entries and the packet's header bits.

The packet processing pipeline will stop when the flow entry that is matched to the packet does not have instructions that request that the packet be sent to another flow table for processing. Once this happens, the rest of the instruction set for this flow table entry will be processed against the packet and then the packet will be forwarded by the switch.

Packets will not always match to the current contents of a flow table. When this occurs, it is called a table-miss. How a table miss is handled is highly dependent on the implementation of the OpenFlow protocol on a switch. Many different actions can be taken including dropping the packet, passing the packet on to another table, or sending it to the controller via the control channel using *packet-in* messages [20].

10.1.2.2 Flow Tables

At the heart of an OpenFlow switch's packet processing pipeline are its flow tables. Each flow table consists of a number of flow entries. Each flow entry consists of six information items. Two of these items are the match fields and the priority. The match fields are the values that are compared against specific fields in a received packet in order to determine if there is a match.

It is possible that multiple flow table entries may match the same packet at the same time. If this occurs, then the flow entry's priority value is used to determine which match will be used to provide the instructions that will be executed against the packet. Figure 47 shows the steps that an OpenFlow switch goes though once a packet has been received.

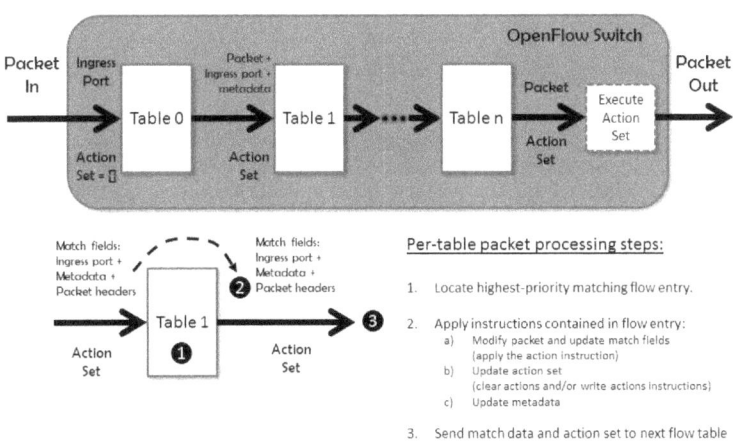

Figure 48: OpenFlow packet flow through the packet processing pipeline [20]

If it exists, when the table miss flow entry is matched to a packet all fields are omitted from the comparison and the table miss flow entry has a priority of 0 (lowest priority).

10.1.2.3 Matching

Figure 48 shows the steps that an OpenFlow switch goes though once a packet has been received.

Every time a flow table entry matches a packet, the counter associated with that flow table entry is updated. These counters are permitted to roll over. After the counters have been updated, then the instructions that are associated with that flow table entry will be executed.

In the unlikely event that two or more flow table entries match the packet and have the same priority level, then the OpenFlow switch will not be able to determine which flow entry's instruction set should be executed. In this case, nobody will win. The OpenFlow switch will assume that the selected flow entry is left undefined and the packet will not be changed.

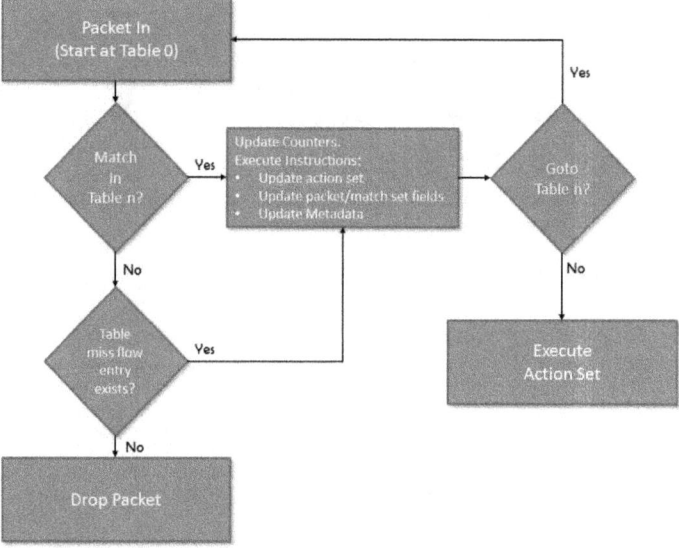

Figure 49: OpenFlow packet matching process

10.1.2.4 Table Miss

A match between a received packet and a flow table entry will only occur if the flow table entry contains values that are the same as found in the header of the packet. In the case that none of the entries in the flow table contain the values that are in the received packet, then a table-miss event has occurred.

The OpenFlow protocol states that every flow table must have a table-miss flow entry. However, the table-miss flow entry does not exist in the flow table by default. This means that the external controller is responsible for creating this table entry.

The table miss entry will match every packet and this match will have a priority of "0". When a packet is unmatched by any other flow table flow entry with a higher priority, the table miss entry will be the only match. When the match with the table miss flow entry is the only match that is made, the table miss flow entry instructions will be applied to the packet.

When a packet does not match any of the other entries in a flow table, the table miss flow entry will determine how to process the packet. The actions that may be taken include dropping the packet, sending the packet to the controller, or sending the packets to another table. There are two actions that the table miss flow entry must support. It must support at least sending packets to the controller using the reserved port CONTROLLER. Additionally, it must support the dropping of packets using the OpenFlow Clear-Actions instruction [20].

10.1.2.5 Flow Removal

A SDN network is dynamic. Links will go up and down all the time. This means that an OpenFlow switch's flow table needs to be constantly updated by the external controller. One key action that will be taken by the controller will be the removal of flows. Removals can occur at the request of the external controller, the switch flow expiration mechanism can cause a flow to be removed, or an optional switch eviction mechanism can be used to remove a flow.

The OpenFlow switch's flow table entry expiration mechanism works in one of two different ways depending on the configuration of the switch. A nonzero hard timeout counter seconds value can be associated with a flow that when enough time has elapsed for it to count down to zero will cause the flow to be removed. Alternatively, a second idle timeout value can be associated with a flow. If this counter has a nonzero seconds value, then if the flow does not match a packet that has been received by the switch before the idle counter reaches zero, then the flow will be removed.

Flows can also be removed by the external controller. The controller accomplishes this removal by sending a message to the switch requesting that the flow be deleted.

Finally, in the case the OpenFlow switch wants to reclaim switch resources, then it can evict one or more flow entries. Note that the ability to evict flows is an optional OpenFlow switch feature and must be enabled before it can be used.

The external controller can configure the switch to make it notify the controller whenever a flow is removed. This setting can be made on a flow-by-flow basis. Every flow removal message contains four pieces of information:

- A complete description of the flow entry
- The reason for the flow entry's removal (expiration, deletion, or eviction)
- The flow entry's duration at the time of its removal
- The flow entry's flow statistics at the time of its removal

10.1.2.6 Meter Table
The OpenFlow protocol provided the ability to implement a simple Quality-of-Service (QoS) operating by using what are

called "meters". Meters can be defined on a per flow basis. The OpenFlow protocol specification does not contain any required meter band types. There are only optional meter band types. In order to implement metering, the switch must allow the meter to measure the rate of the packets that are assigned to it. The goal of this metering is allow the switch to control the rate of those packets that are being metered.

In order to implement the metered QoS features, the OpenFlow switch must first provide counters so that the rate of packets can be counted and measured. Eight types of counters are maintained. Counters are maintained for the following OpenFlow components:

1. Each flow table,
2. Each flow entry,
3. Each port,
4. Each queue,
5. Each group,
6. Each group bucket,
7. Each meter,
8. Each meter band

The counters used by an OpenFlow switch can be implemented in software and then maintained by polling of hardware counters that have a more limited range that they can count. The counters listed in Table 2 are required counters that every OpenFlow switch must have:

Table 2: List of Required Counters

Type	Counter	Bits
Per Flow Table	Reference Count (active entries)	32

Per Flow Entry	Duration (seconds)	32
Per Port	Received Packets	64
Per Port	Transmitted Packets	64
Per Port	Duration (seconds)	32
Per Queue	Transmit Packets	64
Per Queue	Duration (seconds)	32
Per Group	Duration (seconds)	32
Per Meter	Duration (seconds	32

An OpenFlow switch is not required to support all possible counters. The optional counters that an OpenFlow switch can support are shown in Table 3.

For each counter the duration refers to the amount of time that a flow entry, a port, a group, a queue, or a meter has been installed in the OpenFlow switch. Each counter must be tracked with per second precision.

The Receive Errors counter is the total of all receive and collision errors associated with the counters in Table 2 and Table 3 as well as any others that are not identified by these tables.

Each counter is unsigned and will wrap around with no indication that this has happened if its maximum value is exceeded.

Table 3: List of Optional Counters

Type	Counter	Bits
Per Flow Table	Packet Lookups	64
Per Flow Table	Packet Matches	64
Per Flow Entry	Received Packets	64
Per Flow Entry	Received Bytes	64
Per Flow Entry	Duration (nanoseconds)	32
Per Port	Received Bytes	64
Per Port	Transmitted Bytes	64
Per Port	Receive Drops	64
Per Port	Transmit Drops	64
Per Port	Receive Errors	64
Per Port	Transmit Errors	64
Per Port	Receive Frame Alignment Errors	64
Per Port	Receive Overrun Errors	64
Per Port	Receive CRC Errors	64
Per Port	Collisions	64
Per Port	Duration (nanoseconds)	32
Per Port		
Per Queue	Transmit Bytes	64
Per Queue	Transmit Overrun Errors	64
Per Queue	Duration (nanoseconds)	32
Per Group	Reference Count (flow entries)	32
Per Group	Packet Count	64
Per Group	Byte Count	64
Per Group	Duration (nanoseconds)	32

Per Group Bucket	Packet Count	64
Per Group Bucket	Byte Count	64
Per Meter	Flow Count	32
Per Meter	Input Packet Count	64
Per Meter	Input Byte Count	64
Per Meter	Duration (nanoseconds)	32
Per Meter Band	In Band Packet Count	64
Per Meter Band	In Band Byte Count	64

If an OpenFlow switch does not provide a specific numeric counter, then its value must be set to the counter's maximum field value which is the unsigned equivalent of "-1".

10.1.2.7 Instructions

In order for a packet in a SDN network to get from its source to its destination, it will need to have its header modified by the OpenFlow switches that it passes though. The way that an OpenFlow switch determines how to modify a packet is based on the instructions that it executes against the packet. The collection of instructions that get executed will be based on which instructions are contained in the flow table entry that the packet's header fields match to.

If for any reason an OpenFlow switch is unable to execute the instructions associated with a flow entry, then the switch must reject the flow entry. If this event happens, then the switch must return an unsupported flow error message. It is possible that OpenFlow switch flow tables may not support every instruction, every action, or every match.

10.1.2.8 Action Set

In additional to the instructions that are contained in the flow table entries, each packet in an OpenFlow switch has an action set associated with it. Initially, this action set is empty.

Depending on a match that is made with a packet, the corresponding flow entry can modify the packet's action set. The packet's action set travels with it as it moves between flow tables. The actions in the packet's action set will be executed when the instruction set of a flow entry does not contain an instruction that points to the next flow table and the pipeline processing of the packet has stopped.

10.1.2.9 Action List

The list of actions contained in a packet's action set is called an action list. The order in which the actions in the action list are executed is determined by their location within the action list. When they are executed, the results are immediately applied to the packet.

The way that an action list is executed is to start by executing the first action on the list on the packet and then execute each action in the list in sequence on the packet. The effect of executing these actions on the packet is cumulative. A copy of the packet will be forwarded to the designated port in its current state if the action list contains an output action. A copy of the packet in its current state will be processed by the appropriate group buckets if the action list contains group actions.

10.1.3 The OpenFlow Channel

An OpenFlow switch is connected to an external controller via an OpenFlow channel. This is the interface that the controller uses to configure and manage the switch, receive events from the switch, and send packets out of the switch.

The OpenFlow protocol supports the following three types of messages for exchanging information between the controller and the OpenFlow switch:

1. Controller-to-switch,
2. Asynchronous
3. Symmetric

Controller-to-switch messages are used to directly manage or inspect the state of the switch and are initiated by the controller. Asynchronous messages are used to update the controller of network events and changes to the switch state and are initiated by the switch. Symmetric messages are sent without solicitation and are initiated by either the switch or the controller

There are seven Controller-to-switch messages that are initiated by the controller and may or may not require a response from the switch.

The controller does not have to request that asynchronous messages be sent from an OpenFlow switch. Asynchronous messages are sent by the switch to controllers in order to denote a packet arrival or switch state change. There are three main types of asynchronous messages.

Symmetric messages can be sent without solicitation in either direction. Four symmetric messages have been defined as a part of the OpenFlow protocol.

10.1.3.1 Message Handling

The OpenFlow protocol does provide a reliable means for both message delivery and processing. However, the OpenFlow protocol does not automatically provide acknowledgements or ensure ordered message processing.

When a switch receives a message from the controller, it must process it and depending on the type of message the switch may generate a reply. An error message will be sent to the controller if the switch cannot completely process a message that it has received from a controller. It is important that the controller's view of the switch be kept consistent with the state of the OpenFlow switch.

10.1.3.2 OpenFlow Channel Connections

An OpenFlow switch and an OpenFlow controller exchange OpenFlow messages using an OpenFlow channel. In general, a single OpenFlow controller will communicate with multiple OpenFlow switches using multiple OpenFlow channels. A single OpenFlow switch will typically have either a single OpenFlow channel connection to a single OpenFlow controller or multiple OpenFlow connections to multiple OpenFlow controllers for backup and reliability.

An OpenFlow controller is generally located remotely and uses one or more networks to connect to a given OpenFlow switch. The only requirement of the controller / switch network is that it support the TCP / IP protocol. The network that is used to support controller to switch communications can be a dedicated network, a shared network, or an in-band network (the network that is being managed by the OpenFlow switch).

The OpenFlow channel between the OpenFlow switch and the OpenFlow controller is generally a single network connection that uses the Transport Layer Security (TLS) or plain TCP protocol. It is possible to create an OpenFlow connection that is composed of multiple network connections in order to exploit parallelism.

The OpenFlow switch is responsible for establishing a connection with the OpenFlow controller. In some cases, the OpenFlow switch may permit the OpenFlow controller to establish a connection with it. However, in this case, the switch usually should restrict itself to using only secured connections (TLS) in order to prevent unauthorized access to the switch.

The port that is used to establish communication between the OpenFlow switch and the external controller must be user-configurable, but otherwise fixed, IP address. This connection can be established using either a user-specified transport port or the default transport port.

Assuming that the switch has been pre-configured with the IP address of the controller to connect to, the switch will then initiate a standard TLS or TCP connection to the controller. Traffic both to and from the OpenFlow channel does not travel through the packet processing pipeline. What this means is that the OpenFlow switch will need to identify incoming traffic as local before checking it against the flow tables.

A switch can detect that it has lost contact with all of the external controllers that it had been connected to when it detects an echo request timeout, TLS session timeouts, or other disconnections. When this happens, the switch must immediately enter either "fail secure mode" or "fail standalone mode", depending upon the switch implementation and configuration. The difference between the two modes is that in "fail secure mode", the only change to switch behavior is that packets and messages destined to the controllers are dropped. In "fail secure mode" flow entries should continue to expire according to their timeouts.

One way that the switch and the controller may communicate is by using a TLS connection. The switch starts the TLS connection to the external controller when the switch is started. This connection will be located by default on TCP port 6653. In order to authenticate both the switch and the controller, certificates signed by a site-specific private key are exchanged. It is required that each switch must be user-configurable with one certificate for authenticating the controller (the "controller certificate") and the other for authenticating to the controller (the "switch certificate").

Communication between the OpenFlow switch and the controller using a plain TCP connection is permitted. The switch will initiate the TCP connection to the controller upon its startup. This TCP connection is located by default on TCP port 6653. The OpenFlow protocol designers recommend that when a plain TCP connection is used, alternative security measures to prevent eavesdropping, controller impersonation or other attacks on the OpenFlow channel are also used.

10.1.3.3 Controller Modes

The OpenFlow protocol supports three different modes for controllers that are connected to a switch. The roles are as follows:

1. EQUAL
2. SLAVE
3. MASTER

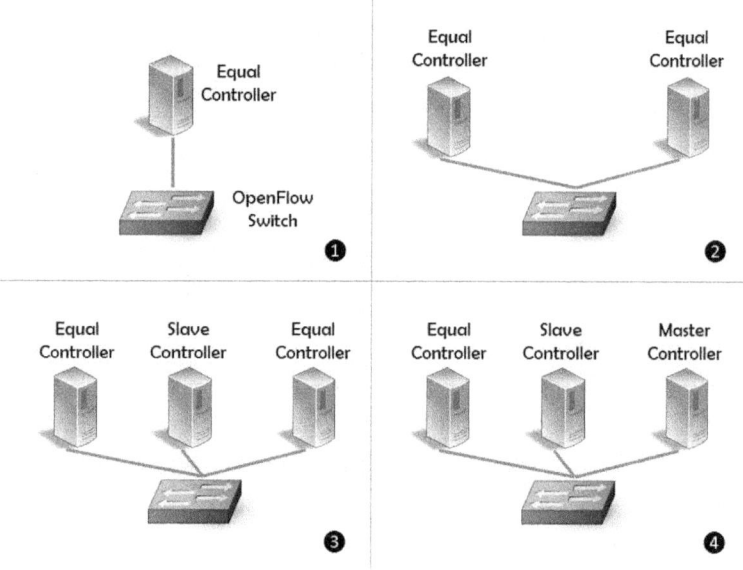

Figure 50: OpenFlow Controller Modes

Figure 49 shows four scenarios for how one or more controllers could connect to an OpenFlow switch. In the first scenario, a single controller connects to the OpenFlow switch. The controller in this scenario is in the EQUAL role. When a controller is in this role, it will have full access to the OpenFlow switch. While in this role the default action is that the controller receives all the switch asynchronous messages. This role permits the controller to send controller-to-switch commands to modify the state of the switch.

In the second scenario, two controllers who are both in the EQUAL role are connected to the same switch. OpenFlow switches can establish communication with a single controller, or they may establish communication with multiple controllers at the same time. If an OpenFlow switch has connections with multiple switches at the same time then they will be considered to be more reliable and can continue to operate in OpenFlow mode if one controller or controller connection fails. When a hand-over between controllers is required, it will be handled by the controllers themselves and there will be no required switch involvement. This provides two advantages: first that it permits fast recovery from failures and secondly that it permits controller load balancing.

In scenario 3, a single OpenFlow switch is shown connected to three controllers: two in EQUAL mode and one in SLAVE mode. A controller has the ability to request that its role be changed. One way that it can cause this to occur is by requesting that its role to be changed to SLAVE. This role only permits the controller to have read-only access to the OpenFlow switch. The default for this role is for the controller to not receive switch asynchronous messages, apart from Port-status messages. The controller is not permitted to execute all controller-to-switch commands that send packets or modify the state of the switch.

Scenario 4 shows an OpenFlow switch with one EQAL, one SLAVE, and one MASTER controller. When a controller is in the MASTER role, the controller will have full access to the OpenFlow switch. The difference between this role and the EQUAL role is that in this role the switch will ensure that this controller is the only controller that is currently in this role. The switch is prohibited from changing the state of a controller on its own.

10.1.3.4 Use Of Auxiliary Connections To Boost Performance And Reliability

When a controller is managing an OpenFlow switch, the performance of the channel between the controller and the switch may become a bottleneck. In order to alleviate this problem, the OpenFlow protocol permits multiple auxiliary connections to be established between the controller and the switch in addition to a main connection.

The OpenFlow channel traditionally consists of a "main connection". In order to boost the performance of the controller to switch communication along with the reliability of the connection, additional auxiliary connections between the controller and the switch can be created by the switch. This is shown in Figure 50.

Each auxiliary connection must use the same controller IP address as the main connection. However, different transport protocols can be used on each auxiliary connection. These protocols can include TLS, TCP, DTLS, UDP, etc.

Prior to establishing any auxiliary connections the switch must first establish its main connection. In the case that something happens to the main connection and it becomes unavailable to the switch, then all of the auxiliary connections that have been set up need to immediately be taken down.

10.1.3.5 Synchronization Of Flow Tables

An OpenFlow switch may contain multiple flow tables as a part of its packet processing pipeline. The switch may decide that a table in this pipeline should be synchronized with another table in the pipeline. What this means is that any additions or deletions to one table will be automatically replicated by the switch to the synchronized table.

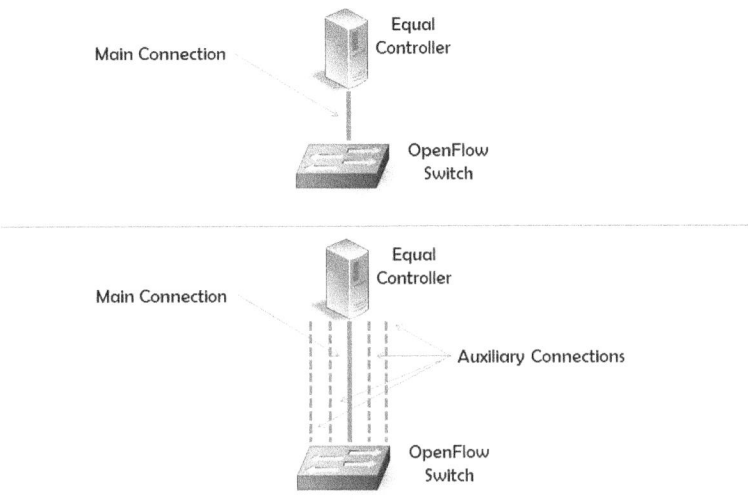

Figure 51: Auxiliary Connections To An OpenFlow Switch

The purpose of this type of synchronization is to permit the packet processing pipeline to potently perform multiple matches on the same packet at different points in the pipeline. Switch implementation and configuration will determine exactly how tables will be synchronized because the OpenFlow protocol does not specify the details of how this is to be done.

10.1.3.6 Bundle Messages

A controller will often have a need to make a large number of modifications to the configuration of an OpenFlow switch. There are two ways to go about making these changes. The first is to send each change message to the OpenFlow switch individually and have the switch make the change. Alternatively, a sequence of OpenFlow modification requests known as a "bundle" can be sent from the controller to the switch as part of a single message.

Bundles are handled uniquely by a switch. If all of the modifications contained in a bundle can successfully be made to a switch, then the modifications will be made and the results will be retained by the switch. However, if for some reason an error is encountered while performing one or more of the modifications, the modifications will not be made and the results will not be retained.

The use of bundles allows a controller to synchronize its changes across multiple OpenFlow switches. Bundles can be sent to multiple switches where they will be prevalidated. Once this has been accomplished for all switches, the changes can then all be applied at the same time.

10.2 OpenFlow Configuration and Management Protocol

A separate configuration and management protocol has been created to manage the communication path that the OpenFlow controller and OpenFlow switch use to communicate. This protocol is called OF-CONFIG. The purpose of OF-CONFIG is to provide functionality that is not found in the OpenFlow protocol. The creation of the OpenFlow switch is assumed to have been done independently of the OpenFlow and OF-CONFIG protocols.

The OF-CONFIG protocol can be used to remotely configure the OpenFlow switch's configuration and operation of the datapath that connects the switch to one or more controllers. The OP-CONFIG identifies the various types (physical and virtual) of

OpenFlow switches as OpenFlow Logical Switches. The components of the OF-CONFIG protocol are shown in Figure 51.

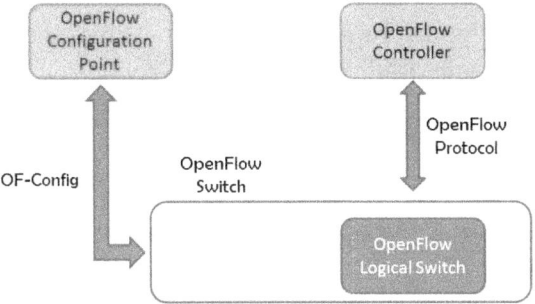

Figure 52: Relationship between OF-CONFIG protocol and the OpenFlow protocol

The OF-CONFIG protocol provides the ability to perform six functions:

1. The assignment of one or more OpenFlow controllers to interact with a given OpenFlow switch.

2. The configuration of the queues and ports that an OpenFlow switch will use to communicate with one or more controllers.

3. The ability to remotely change some aspects of an OpenFlow switch's ports (e.g. place them in an up/down state).

4. Configuration of security certificates for secure communication between the OpenFlow Logical Switches and OpenFlow Controllers

5. Discovery of capabilities supported by an OpenFlow Logical Switch

6. Configuration of a small set of tunnel types such as such as IP-in-GRE, NV-GRE, and VxLAN [22].

OP-CONFIG messages to the OpenFlow switch come from an OpenFlow Configuration Point. The OF-CONFIG specification does not detail exactly what an OpenFlow Configuration Point is.

10.2.1 Setting Up A Connection Between A Switch And A Controller

The OpenFlow protocol specifies that the OpenFlow switch set up the connection between it and the controller. In order to perform this action, the OP-CONFIG's OpenFlow Configuration Point has to have configured the OpenFlow switch with the following three informational components:

1. The controller's IP address

2. The controller's port number

3. The transport protocol to be used by the connection (TLS or TCP)

Network devices that support the OF-CONFIG need to implement the NETCONF protocol to be used as the transport protocol for OF-CONFIG.

The OF-CONFIG protocol provides the following functionality:

- The OF-CONFIG protocol specifies how to configure multiple instances of the parameter set for specifying the connection setup to multiple controllers.

- When a switch loses connectivity with a controller, it has to enter into the fail secure mode or the fail standalone mode. The OF-CONFIG protocol is responsible for configuring the parameters that the switch uses to determine which mode to enter when connection is lost.

- An OpenFlow switch has a number of queues that are used to process packets. Each queue has three parameters associated with it: min-rate, max-rate, and experimenter. The OF-CONFIG protocol is responsible for providing the information that is needed to configure these parameters for each queue.

- An OpenFlow switch has a number of ports that are used to communicate with other OpenFlow switches and controllers. OF-CONFIG is responsible for configuring four port related parameters: no-receive, no-forward, no-packet-in, and admin-state. OF-CONFIG also supports providing configuration information for port features and for getting state information on these port features.

- Every OpenFlow switch has a set of unique capabilities. The OF-CONFIG protocol provides a means for the switch to inform a controller what capabilities have been implemented on it.

- Each connection from the switch to the controller is identified by the switch Datapath ID and an Auxiliary ID.

The OF-CONFIG protocol provides a means to configure the Datapath ID.

10.2.1.1 OF-CONFIG Transport Protocol

In order for an OpenFlow switch to support the OF-CONFIG protocol, it must support the Network Configuration Protocol (NETCONF). This is an IETF network management protocol.

The switch's support of the NETCONF protocol provides the OpenFlow Configuration Point with the mechanisms to install, manipulate, and delete the configuration of OpenFlow switches. Its operations are implemented using a simple Remote Procedure Call (RPC) layer. The NETCONF protocol uses an Extensible Markup Language (XML) based data encoding for the configuration data as well as for its protocol messages. The protocol messages are then exchanged on top of a secure transport protocol.

10.3 The Conformance Test Specification for OpenFlow Switch Specification 1.0.1

The Open Networking Foundation has created a set of 206 test cases that can be used to determine the conformance of an OpenFlow 1.0.1 enabled switch to the OpenFlow protocol specification (v1.0.0 and the subsequent Errata v1.0.1). Currently no test procedures have been created to validate security, interoperability or performance features of an OpenFlow switch.

Some of the test cases contained in this test specification are mutually exclusive, optional or only relevant for some implementations. In some of the test cases, the methods of validating the outcome of tests are not fully described and may be left up to the tester or test tool developer.

Three different profiles have been created that OpenFlow switches can conform to. These profiles are as follows:

1. Full
2. L2
3. L3

If a switch is to be considered fully conformant with the OpenFlow Switch Specification 1.0.0 and the subsequent OpenFlow Switch Errata 1.0.1, then for a given profile the switch must satisfy the requirements of all test cases that indicate "MANDATORY" for "All" or the appropriate profile (Full / L2 / L3).

The test bed used to perform the test cases is shown in Figure 52. The test bed that is used to perform the tests consists of the following components:

1. A single test controller that has a single control channel connection to the device under test (DUT).

2. The test controller should be equipped with the ability to perform a packet trace and decode OpenFlow 1.0 packets.

3. A traffic generator/analyzer that has a minimum of 4 ports compatible with the device under test for data plane connections.

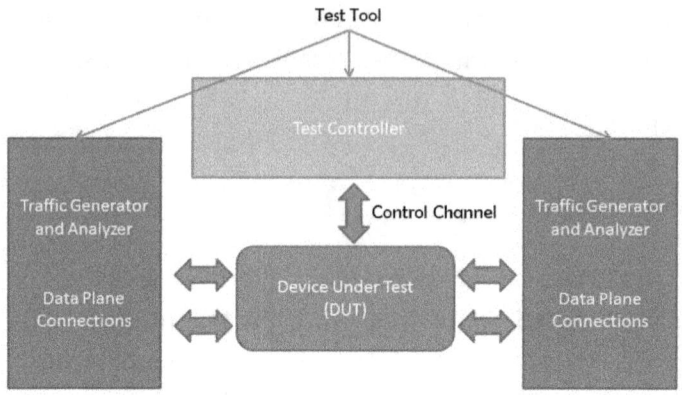

Figure 53: Test bed used to execute OpenFlow test cases

There are 10 groups of tests that have been specified to be performed in order to determine if a device is compliant with the OpenFlow protocol specification [22]. The 10 groups are as follows:

1. BASIC SANITY CHECKS
2. BASIC OPENFLOW PROTOCOL MESSAGES
3. SPANNING TREE
4. FLOW MODIFICATION MESSAGES
5. FLOW MATCHING
6. COUNTERS
7. ACTIONS
8. MESSAGES
9. ASYNC MESSAGES
10. ERROR MESSAGES

10.4 The OpenFlow™ Conformance Testing Program

The Open Networking Foundation (ONF) has established an OpenFlow Conformance Testing Program to assist vendors in certifying that their hardware and software products meet the OpenFlow protocol specifications. The testing that is done as a part of this program is performed by accredited testing labs around the world.

When a vendor's product is able to pass the tests in this program, the product will receive a certificate of conformance. Additionally, the vendor will then be permitted to use the ONF OpenFlow Conformance Testing Program logo as a part of their product marketing efforts. Additionally, vendor's whose products have been certified as a part of this program will be listed on the ONF's web site.

The ONF OpenFlow Conformance certification means that a vendor's product conforms to a specific version of the OpenFlow specification. Currently the ONF offers three different types of certificates: Full Conformance, Layer 3 Conformance, and Layer 2 Conformance. The differences between these certificates are as follows [23]:

- **Full Conformance Certification**: The tested device must be able to match all 12 "match fields" listed in the OpenFlow Switch Specification 1.0.0 (Errata 1.0.1).

- **Layer 3 Conformance**: The tested device must be able to match the following 4 fields in the OpenFlow Switch Specification 1.0.0 (Errata 1.0.1): Ingress Port, Ethernet

Type, IP Source Address and IP Destination address.

- **Layer 2 Conformance**: The tested device must be able to match the following 5 fields in the OpenFlow Switch Specification 1.0.0 (Errata 1.0.1): Ingress Port, Ethernet Source Address, Ethernet Destination Address, Ethernet Type and VLAN id.

In order for a vendor to obtain ONF OpenFlow Conformance certification for their device, they must first join the ONF. The next step is for the vendor to establish a contract with an ONF-Approved Testing Lab. That lab will then conduct a certified conformance test and, assuming that the tested product passes the tests, will certify it. Once this has been accomplished, the ONF will approve the awarding of an ONF OpenFlow Certificate of Conformance.

There are currently three ONF Authorized Conformance Testing Labs:

1. Indiana Center for Network Translational Research and Education (InCNTRE)
   Indianapolis, Indiana, USA

2. Beijing Internet Institute (BII)
   China

3. University of New Hampshire Interoperability Lab (UNH-IOL)
   Durham, New Hampshire, USA

11 The Future Of SDN

Software Defined Networks (SDN) are brand new. The technology has only just been invented and there are very few real-world networks in existence that currently use this technology. This means that there remains a great deal to learn about how to both build and manage SDN networks.

Researchers are using the information that is currently available about the OpenFlow protocol to create test networks and to run network simulations. Of great interest is how SDN and legacy networks will interact. Everyone realizes that there will never be a way to perform a "flash cut" to convert a legacy network into a SDN network overnight and so interoperability is going to be very important.

The initial work on SDN has been to apply it to traditional L2/L3 enterprise networks. Researchers are only now starting to try to determine if SDN can be applied to other types of networks such as optical transport and wireless. If it is possible to extend SDN to cover these additional networking technologies, then the next question that will need to be answered will be to determine how that type of SDN network can be managed.

SDN comes with a number of potential advantages over traditional networking technologies. However, how to maximize these benefits is still unclear. Additional research is going to be required in order to better understand how scalable this approach to networking is.

In a network, a middlebox is a computer networking device that transforms, inspects, filters, or otherwise manipulates traffic for purposes other than packet forwarding. The arrival of SDN networks means that a great deal of current middlebox functionality may be able to be worked into the centralized controller. How to do this and how to accomplish it without degrading the performance of the network are unanswered questions.

In this section we will be taking a look at a number of the open issues in SDN and how researchers are going about investigating them. There are no solid answers as of yet, but the research is ongoing.

11.1 How Can SDN Be Added To Existing Enterprise Networks?

This research tackles the challenging problem of just exactly how new SDN technology can be added to existing enterprise networks. A great deal of the research into SDN networks has assumed that an end-to-end SDN network could be built from the ground up. However, in the real world this will not be possible. Existing legacy networks are going to have be upgraded to use SDN technology. Exactly how to go about doing this has been the subject of ongoing research. [30]

It is acknowledged that once a commitment has been made to transform an existing legacy enterprise network into a SDN network, the transformation will cause a hybrid network to be created. A portion of the network will consist of legacy networking equipment and a portion will consist of new SDN equipment. Researchers want to understand if and how the two different networks can be made to work together.

In this networking scenario, three key questions that need to be answered have been identified:

1. Is there any benefit to the network's owner to upgrade their existing legacy network to become a hybrid Legacy / SDN network?

2. In a hybrid legacy / SDN network, how important is the actual placement of those switches that have been upgraded to SDN?

3. In order to get the greatest benefit from introducing SDN into the enterprise network, just how many of the existing legacy switches need to be upgraded to use SDN technology before benefits will start to be seen?

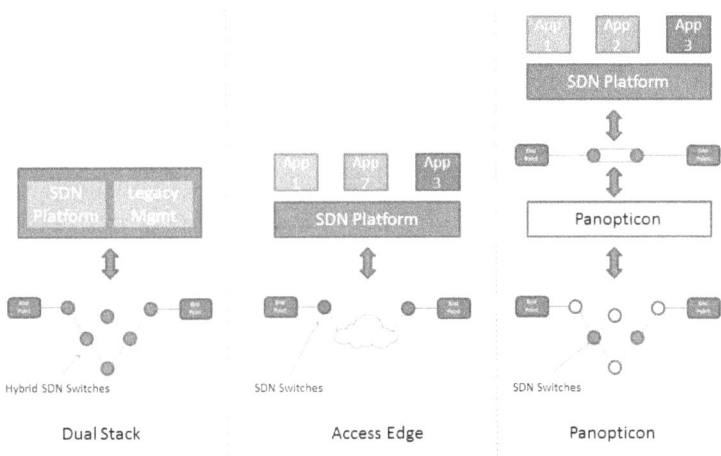

Figure 54: 3 Possible Enterprise Network SDN Deployment Scenarios

Figure 53 shows three possible ways that SDN equipment could be deployed into an existing SDN enterprise network [30]. The dual stack scenario in this figure shows SDN functionality (perhaps support for the OpenFlow protocol) being added to the legacy switches in the network. The result of this addition is that two separate side-by-side networks are created. When applications are connected over the network a decision has to be made whether to send their packets via a SDN or a via a legacy connection. A drawback to this approach is that it will be necessary to continue to deploy more and more switches with both SDN and legacy functionality in order to provide maximum value to the applications that choose to use both forms of connections.

The Access Edge scenario in Figure 53 shows an alternative way of introducing SDN technology into a preexisting network. In this scenario, the SDN functionality is only introduced at the edge of the network – in the access network. The core network will remain the domain of the legacy network switching gear. This approach would allow entire datacenters that were part of the enterprise network to be virtualized using SDN technology. The SDN network will meet the legacy network when a SDN switch is interfaced to a legacy access switch. In order for the legacy access switch to be able to speak to the SDN switch, its software will have to be updated. In a typical enterprise network there may be hundreds if not thousands of these access switches that will need to be updated. Clearly the expense of implementing this solution will be great.

The final scenario show in Figure 53 is called Panopticon. In this scenario the SDN and legacy networks are integrated together. The integrated network is what is shown to the SDN network controller. The researchers report that from this scenario they have discovered that "... *the key benefits of the SDN paradigm can be realized for every source-destination path that includes at least one SDN switch.*" [30]

The conclusion that the researchers have reached is that not every switch in the enterprise network has to be SDN switch in order to get the value of deploying SDN technology. Instead, only a subset of the switches need to be able to support SDN. As long as every path goes through at least one SDN switch, then a programmable network access control policy can be implemented. Additional functionality, such as load balancing, can be implemented for those network connections that pass through two or more SDN switches.

The Panopticon network architecture that the researchers have created ensures that every network connection will pass through at least one SDN node. The Panopticon architecture uses network waypoints to control network traffic and thereby implements the SDN abstraction.

A prototype of the Panopticon network architecture has been implemented and many simulations of the network have been run. The result of these studies have revealed that only an astonishing 0.6% of the switches in a typical enterprise network have to be upgraded before roughly 80% of the network can be operated as a single SDN network. Note that in this network, VLAN and flow table updates can successfully be met.

Since the Panopticon network architecture allows the entire network to be viewed by the SDN controller as a single big virtual switch, end-to-end policies can be implemented on it. These types of policies include such things as application load balancing and access control.

One of the most interesting results that has come out of the research on the Panopticon network architecture is the fact that all of the switches in a network do not have to be upgraded to support SDN in order for the company to realize the benefits of SDN. What the Panopticon has revealed is that a partial deployment of SDN may be all that some enterprise networks need in order to realize the SDN functionality that they want while living within the real-world constraints of limited network budgets and other resources.

11.2 Can SDN Be Used With Transport Networks?

Researchers have studied the prices of the three major groups of equipment that are used in networks: processing and storage, routing and switching, and transport. They have discovered that while processing and storage costs are dropping the fastest, transport costs are dropping the slowest. The reasons for this are varied, but they include the fact that they are complex systems that deal with high bandwidth communications and there simply are fewer transport systems that are sold on average. What this means is that any changes to how a network is constructed that allows the existing transport network to be better utilized will provide significant benefits to the network operator.

The service providers who are in charge of transport networks are highly motivated to make their transport networks more sensitive to the changing bandwidth needs of the applications that are using them. Additionally, it has been noted that during weekday business hours business users constitute the majority of transport network users. However, at night and on weekends (or holidays) residential users make up the majority of transport network users. The ability to reconfigure the transport network in order to better meet the needs of these two very different groups of users is an important goal for service providers.

The arrival of virtualization services as described in this book means that companies will soon have the ability to make use of external 3rd party providers of virtual private network (VPN) and cloud-based services. In order to make use of these services, transport facilities that connect the enterprise network to the external network will be required. The amount of bandwidth that will be required for these connections will vary and needs to be adjustable to meet current needs.

The ultimate goal is to be able to provide applications with the ability to specify both the bandwidth and the quality of service (QoS) parameters that they will require from the network. This request will need to be honored by the processing and storage components, the routing and switching network, and the transport network.

In trying to understand the need for introducing SDN technology into the transport network, researchers have studied several different scenarios in which transport networks are used. One such scenario involves the use of "cloudbursting".

Cloudbursting occurs when an enterprise network elects to use external computing resources to augment its own computing power. This could occur during a busy holiday period or perhaps when yearend tax and other filings have to be created. Cloudbursting requires that the external network be initially configured to perform the tasks that have to be done. This generally requires that a great deal of data be transported from the enterprise network to the cloud based network. This data can be customer related data, virtual machines, etc. [31]

The researchers believe that when smaller connections are being used to transport data (< 1Gbps) packet switched networks will be able to efficiently handle the task. However, when the data transportation needs grow to become very large as during a Cloudbursting event (> 10 Gbps) a better method is required. This method could be to bypass the packet network and directly use the transport network to connect the enterprise data center to the external data center. An OpenFlow based SDN can be used to implement this type of a solution. The goal will be to improve the efficiency of the transport network and to avoid any occurrence of blocking in the network.

The researches acknowledge that it may be necessary to perform some rearranging of the existing transport network in order to permit it to support and work with the SDN network. Ultimately, the SDN controller will have to work with both the SDN components of the network and the legacy packet and transport network elements.

One advantage of incorporating the transport network into the overall network SDN processing is that improvements in network efficiency can be achieved. The routing and switching components of the network will be rapidly reconfigured in order to support the changing needs of applications. It is anticipated that the transport portion of the network would be reconfigured at a much slower rate so as to not cause packet reordering and potentially packet loss. [31]

One of the differences between the networks that use transport elements and other enterprise networks is that the transport networks often extend over multiple domains. The researchers point out that current interdomain routing and signaling standards do not currently support the ability to reconfigure the transport network to support the type of end-to-end optimization that we have been discussing [31]. This is an area of networking that still requires more work.

The researchers agree that the arrival of SDN will change networking forever. This is going to mean that how transport networks are configured and used will also be changing. The use of the OpenFlow protocol will allow an abstraction of the physical network to be constructed and will ultimately result in greater simplicity when it comes to managing and configuring the network. In order to incorporate the transport networks that are already a part of today's enterprise networks into this new SDN future, changes are going to have to be made.

The researchers believe that the focus of future research needs to be on how best to provide bandwidth on demand features as multilayer networks are integrated with cloud networks. A great deal of research has already been put into applying SDN technology to switching networks. Now the researchers believe work needs to done in order to determine how to use OpenFlow and controllers to optimize how the transport network supports application's bandwidth needs.

11.3 How Can SDN Concepts Be Extended To Work With Optical Transport Networks?

Researchers agree that in an ideal world, applications could be completely independent of the networks that they use to communicate. However, because of real-world communications issues such as latency, bit rate, and packet loss, applications have to be able to ensure that the network will be able to meet the application's needs.

Applications need to be able to communicate with the network's control plane in order to establish the types of network routes and to identify the network resources that will meet the application's communication needs. Transport networks represent a special case when it comes to the control plane. Due to the distances that they cover, transport networks often have multiple segments, each of which may have its own technology and administration. In order to provide applications with the services that they need, interoperability between these various segments will be required.

Additionally, in today's transport networks, many of the administrative functions are performed manually. Researchers hope that the arrival of SDN may be able to automate many of these functions [31]. The hope is that many of today's manual processes which can be time consuming and difficult to perform can now be automated in order to optimize how network resources are being used.

The OpenFlow protocol does not currently provide the ability to configure wavelength or circuit based equipment. Today's transport networks are generally controlled by Element Management Systems (EMS) or Network Management Systems (NMS) which are designed to permit manual control and configuration of the transport network. These management systems can be enhanced by adding a programmable interface (API) that will permit them to be externally controlled by another application. It is through an interface like this that the SDN control plane may be able to control an optical transport network.

One of the biggest challenges that transport networks face is that their connections can have very specific performance characteristics. These characteristics can include, but are not limited to, bandwidth, connectivity, quality-of-service (QoS), and resiliency. These parameters can change over a period of time, such as workday requirements being different from night time requirements. If there was a way to dynamically allocate transport network resources then the network operator could realize improved network utilization.

There are many different ways to architect an optical transport network. Each configuration can potentially be managed as part of a SDN network. Within a typical transport network, the switching can be performed based on packets, time slots, or fiber. Wavelength switching is a relatively new form of transport network. Wavelengths ("colors") paths can be switched; however, the process is slow and can take several minutes to complete.

Within a transport network, the control plane can either be distributed or centralized. In the case in which a single network end point is to be connected to multiple network end points, it is possible that a connection can be set up; however, it may not be an optimal connection. Links may be over or under loaded, paths may be too loaded to accept another link, or a connection cannot be made because the network topology is not known.

Many of these problems can be solved by using a centralized controller. In a transport network, a centralized controller has the ability to compute an end-to-end path that satisfies an application's connection, circuit, or wavelength constraints [31]. As in packet SDN networks, a centralized controller in a transport network is better able to create more efficient links. When a path is needed as part of a high priority connection, existing connections on that path can be rerouted to other paths and the final network configuration can still meet the performance needs of all connections involved. Diverse routes for connections that share the same set of end points can also easily be calculated.

Researchers have noted that a centralized controller may not always be the ideal solution for a transport network. It is believed that a centralized controller may be slow to react to a link outage. In this case, a distributed controller that was running on the same processing platform as the fiber switch would be able to react quicker and minimize packet loss.

The scale of transport networks makes them look different from most packet networks. Researchers believe that this is going to require a different architecture for the control plane that will be used with transport networks. A single unified controller will probably be made up of a series of geographically distributed controllers. These controllers will be responsible for managing their portion of the transport network as well as for communicating with the other controllers. Taken together, the multiple controllers will create a single unified controller for the entire transport network.

Transport networks often cross service provider and geographic boundaries. This means that different parties will be responsible for controlling different parts of the same transport network. Interfaces between the different controllers will be required in order to provide applications with end-to-end connection management. Researchers believe that this can be handled by using simple APIs or existing mature protocols [31].

Researchers believe that just one type of controller will not be enough for transport networks. Instead, both a centralized and a distributed controller functionality will be required. The centralized controller will be responsible for coordinating an application's requirements with the network while the distributed controller will be responsible for providing fast reaction to network events such as a link failure. It is believed that the embedded controller may be more scalable and can be built on existing mature network protocols.

Transport networks are one part of a multilayer network. Adding a SDN controller to such a network provides an opportunity for the optical transport network to be better utilized in order to connect network applications. With the appropriate programming, a SDN controller could configure the optical transport network by setting the modulation scheme, symbol rate, FEC overhead, etc. [31]. Proper control of the transport network could require that the controller have knowledge of vendor specific equipment characteristics and because of this a two-tier controller structure, centralized and distributed, may be the optimal design. Researchers believe that the two separate controllers may operate on different timescales.

11.4 Can A WAN Network's Utilization Be Increased By Using SDN Technology?

The number of data centers is increasing and their importance is growing at the same time. Many of the world's largest service providers, Google, Amazon, Microsoft, IBM, etc. use a large number of data centers as a key part of their service offerings. These firms have constructed separate WANs to interconnect these data centers. Google refers to its inter-data center WAN as the G-Scale backbone network.

These WANs are expensive to build and to operate. The incredible bandwidth that is required comes with an equivalent large price tag. Unfortunately, studies have shown that the average utilization of the links in these expensive WAN networks is generally 40% - 60% at best [33]. The reasons for this low utilization are many and varied, but the fact that applications are unable to communicate when they need the network and how much bandwidth they will need over time results in each connection reserving the maximum amount of bandwidth that it will need for the entire duration of its connection.

One of the key reasons for the poor utilization of WAN links is due to the fact that no global view of the network exists. In the absence of this view, routers will make decisions that are optimal for their local network environment. The end result will be a suboptimal overall network configuration.

Researchers have created the Software-Driven WAN (SWAN). This is a system that has been designed to allow a WAN to better utilize its links by using SDN technology along with a controller that maintains a global view of the network.

Any SDN network that wants to provide high utilization has to perform rapid updates of each switch's flow tables as the needs of the network applications change. However, this introduces problems into the network. As the updates are made, transient network conditions can be created. The changes have to be made to multiple switches in the network. The time that exists between when one switch has been updated and when the other switches have been updated allows network congestion to occur. This congestion can then have a negative impact on any applications that are latency sensitive.

The SWAN system addresses the update / congestion problem by reserving a portion of every path in the network (roughly 10%) for use by it to provide switch update information ("scratch capacity") in a non-blocking fashion. Another challenge that the researchers encountered was that the number of forwarding rules that a given commodity SDN switch would need to support would far exceed its forwarding rule capacity. The researchers addressed this issue by creating a system that would dynamically change a switch's forwarding rules based on traffic demand.

Not all traffic in a data center environment is the same. When researchers talked with service operators, they discovered that there were three primary types of services that a data center WAN would need to support [33]:

1. **Interactive**: an interactive service is a request by one data center to another data center in order to obtain information that is needed to provide an end user with a reply. Characteristics of this type of data center traffic is that it is very sensitive to both loss and delay.

2. **Elastic**: an elastic service is one in which two data centers exchange information in order to ensure that accurate answers can be provided to queries that are sent to an application in a given data center. This data has to arrive at the other data center within seconds or minutes. The impact of not getting the data there fast enough depends on the application that is using the data.

3. **Background**: a background activity is part of a data center's maintenance and provisioning tasks. This can include copying large data sets between data centers for back up or to start up an application at a different site. This traffic requires a large amount of bandwidth and does not have a fixed time by which it must

complete; however, completing it as quickly as possible is desired in order to minimize costs.

The amount of data of each type that data centers exchange is not equally distributed. Interactive data is the smallest amount, followed by Elastic, which is then followed by Background in size.

In today's data center networks, the traffic is engineered using existing network protocols such as MPLS-TE. This protocol creates a set of tunnels between two points in the network and then the traffic between those two points is spread over the various tunnels. The priority of the data is determined by assigning a priority to a tunnel and then mapping traffic with that priority to that tunnel. Additionally, each packet contains a differentiated service packet code (DSPC). This can be used to map a packet into a priority queue.

The MPLS system has two problems associated with it in real-world networks: efficiency and sharing. Poor link efficiency is caused by applications in networks not being aware of the bandwidth needs of other applications. Applications request the maximum amount of bandwidth that they'll need independent of any other application's needs at the time. These types of requests can lead to network over and under subscription. Additionally, since a router using the MPLS protocol can only "see" its local environment, it will make locally greedy decisions when trying to create paths. These local solutions may turn out to be inefficient.

Not all services that operate in a data center environment are equal. This means that some services should take priority when requesting network resources. Researchers have discovered that applications tend to obtain throughput based on the rate that they are sending packets [33]. This is a non-ideal outcome. Attempts to map services into queues with their shared priority often does not work out because there simply are not enough queues to support all of the services.

The SWAN system has been designed to preform two functions: carry more traffic and support network wide sharing of networking resources. SWAN supports three types of traffic priority classes: Interactive, Elastic, and Background. In order to permit congestion causing updates of the routers in the network to occur, SWAN uses a multistep update process. This allows the SWAN system to guarantee that non-background traffic will experience no congestion and background traffic will only encounter congestion that can be bounded.

Researchers have discovered that one of the big limitations of SDN routers is that in real world networks they are required to maintain a large number of forwarding rules (perhaps thousands). Most SDN switches do not provide support for that many routing rules. The researchers have devised a method as a part of SWAN to dynamically change the rules and ensure that only the rules that are currently being used are in the switch at any moment in time.

The SWAN architecture consists of three main building blocks: service brokers, network agents, and a controller. The controller is responsible for coordinating all activity in the network. In the network, those service that are *not* interactive will be assigned a service broker who responsibility it will be to aggregate bandwidth demands from host applications and then provide then with the bandwidth that they need. The network agents exist between the controller and the network switches. The SWAN network architecture is show in Figure 54. Here are the details on the various roles that each component plays in the SWAN architecture:

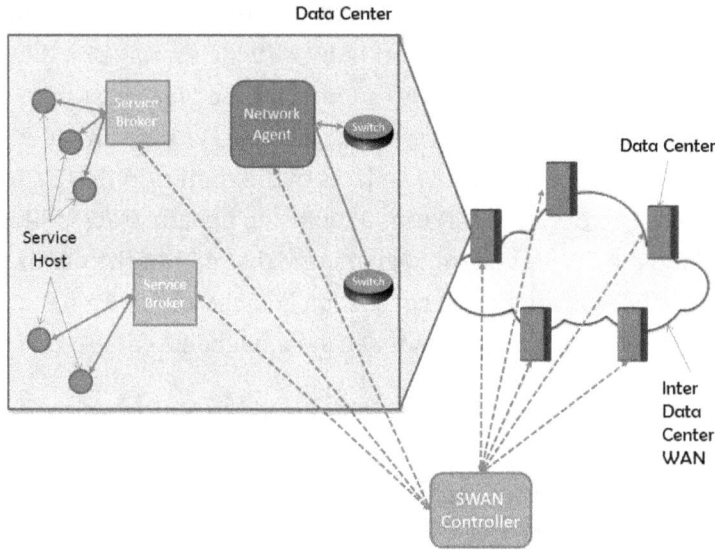

Figure 55: Architecture of SWAN [33]

Service hosts and service brokers: The service brokers have the responsibility of estimating the bandwidth needed by the various service hosts and then limiting it to the maximum bandwidth that the controller has allocated to the service broker. Each of the service hosts makes a request for the amount of bandwidth that will be required for the next 10 seconds. The Service Broker updates the controller with the aggregated demand from the service hosts that it is connected to every 5 minutes.

Network Agents: With help from the network switches, the network agents are responsible for tracking changes in the network's topology and its traffic. Changes in the network's topology are reported to the controller immediately while traffic measurements are reported to the controller every 5 minutes. The network agent has the job of making sure that the OpenFlow flow tables in each of the switches that it is responsible for are correctly updated.

Controller: This network component has the responsibility of performing three tasks every 5 minutes:

1. Calculate the amount of bandwidth that should be allocated to the various network components. This includes each of the service brokers and to the interactive applications.

2. Notify services whose bandwidth allocations have been decreased. Once this task has been performed, the controller will wait a fixed amount of time for the other party to implement the decreased bandwidth usage.

3. Change forwarding states of switches and then communicate to services whose bandwidth has been increased.

One of the most important features of the SWAN architecture is the ability to make many network updates without introducing congestion into the network. In order to make this happen, care is taken when performing two types of updates: traffic distribution across tunnels and when updating tunnels themselves.

When the controller is making updates to how it distributes network traffic across the existing tunnels, it can easily run into a congestion situation when links are heavily loaded. However, because the SWAN architecture requires that some bandwidth be reserved on each link ("scratch bandwidth"), this bandwidth can now be used as a part of the traffic distribution. A series of configuration changes are calculated in order to move the network configuration from where it is to the desired end state. These states are then implemented in each switch using the scratch bandwidth in order to prevent any network congestion from occurring.

Likewise, when updating network tunnels the controller calculates a sequence of changes that can be made to get from the current network configuration to the new network configuration without causing any congestion. The controller then calculates how much traffic from each service can be carried by that tunnel. Next it signals to the services the rate that they are permitted to transmit at. After the services have changed the rate that they are transmitting at, the controller starts to distribute its tunnel changes.

The SWAN architecture has been designed to be able to gracefully handle network failure conditions. The network agents which interface directly with the switches are responsible for detecting switch failure conditions. When a switch or link failure is detected, the network agent will communicate it to the SWAN controller. Upon receiving notification of a failure, the controller will then compute new allocations for the network. All three major components of a SWAN network, service broker, network agent, and controller have backups. In the event that one of these components fails, the backup will take over. The backups are not "hot" backups – they do not maintain a copy of the current state of the network. Instead, upon taking over, they will query the other network components to build a map of the current state of the devices that are attached to them.

In order to study the level of optimization that the SWAN architecture could deliver, the researchers constructed a test bed. Their test bed network consisted of five data centers that occupied three different contents. Each one of the data centers had two WAN facing switches and five servers per data center [33]. The researchers compared the SWAN architecture to that of a MPLS-TE solution. The MPLS-TE solution was able to carry approximately 60% of the offered traffic. The SWAN architecture, however, was able to carry 98% of the traffic. This means that the SWAN architecture was able to carry 60% more traffic than the MPLS-TE solution. Ultimately, this was the goal that the SWAN researchers had designed the SWAN solution to accomplish.

11.5 How Scalable Are Software Defined Networks?

SDN networks permit the control plane to be separated from the data plane. This allows both of the sets of devices in these two planes to now evolve separately. However, SDN comes with a unique set of concerns. With a single control plane for an entire network, important questions about the control plane's scalability and its performance have arisen. Test bed networks are a good way to test and evaluate protocol features; however, what will happen when SDN is applied to very large real-world networks?

11.5.1 SDN Scalability Issues

Networks grow in many different ways. Switches are added, bandwidth increases, and more flows can be handled. A centralized controller is not believed to be able to grow at the same rate as the network. This leads to concerns about the controller becoming a bottleneck in the network. This concern was heightened by early test results that were based on the first SDN controller, NOX. It was reported that the NOX controller could only process 30,000 flow initiations per second [34]. Note that these flow initiations had to support a less than 10ms flow installation. The concern in the SDN community was that 30,000 was the limit for the size of a SDN network.

11.5.2 Controller Designs For Scalability

One SDN network configuration consists of having a single SDN controller provide service for the entire network. Clearly as the network grows, the potential arises that the controller can be overwhelmed with network events and bandwidth requests. No matter how powerful the server that the controller is running on is, there will eventually come a point when the controller is not able to meet the needs of the network. Researchers point out that a data center environment may quickly reach this point with a very high flow initiation request rate.

In order to improve the controller's ability to scale with the network, there are three changes to the controller that can be made. The first is to optimize the I/O processing of the controller. This can include adding I/O batching to minimize the overhead of I/O, porting the I/O handling harness to the Boost Asynchronous I/O (ASIO) library (which simplifies multi-threaded operation). Additionally, making the controller code aware of the environment in which it is running by using a fast multiprocessor-aware malloc implementation that scales well in a multi-core machine [35].

Next, the load on the controller can be minimized by reducing the number of requests that the network forwards to the controller. One way to do this is to have smaller flows handled by custom hardware and only send the larger flows to the controller for processing. This reduces the load on the controller; however, fine-grained flow-level visibility is surrendered in order to gain controller scalability.

Finally, the controller itself can be distributed over multiple systems. Within an SDN network, several servers can fulfill the role of controller. What is required is that there is a single unified view of the network. However, it should be noted that the more consistent the multiple controllers attempt to be, the greater the delay that will be introduced into the system will be. Relaxing the requirement that each of the multiple controllers has to be consistent with the other controllers at all times will reduce the overhead that is required to make this happen.

An alternative approach to distributing the controller consists not of multiple controllers, but rather of a two-part controller. One part is a locally scoped application which is deployed close to the switch. This application is then responsible for processing frequent requests that are local in scope. By doing this, a portion of the processing load that would be put on the central controller is removed. This solution also has a centralized controller. The centralized controller handles any requests that are network wide while at the same time helping to arbitrate between locally scoped applications.

Researchers believe that the scalability issues that are faced by SDN are not unique to SDN [34]. Both the convergence and consistency issues that SDN faces are found in other protocols.

However, what makes SDN different are the following two observations:

1. SDN allows the network designer to make their own set of design trade-offs based on the constraints imposed on the SDN control program design.

2. Software development and standard distributed system design techniques can be used when designing a SDN controller. This will allow standard software development practices to be further used to verify and debug the controller design.

SDN frees the network designer from issues that a traditional network protocol designer has to repeatedly resolve. These issues include resiliency, distributing state information to parts of the network, and discovering the topology of the network. The availability of this information makes it easier to create applications that will eventually run on the controller platform.

11.5.3 Potential SDN Network Scalability Issues

In SDN networks there are two types of flows: reactive and proactive. Reactive flows are created when a switch receives a packet that it currently does not have a flow table entry that will process it. When this happens, the switch reaches out to the controller, the controller provides the switch with an update to its flow table with which to handle the packet and then the packet is processed. This approach allows for fine-grained high-level network wide policy enforcement.

In a proactive scenario the controller updates the switch with the new flow table information before the packet is received.

Proactive flow setup designs allow the controller to avoid any flow-setup delay penalty.

In an SDN network, the flow setup process has four steps [34] which are shown in Figure 55:

1. A SDN network switch receives a packet. Upon checking the packet's header information, the switch determines that it does not match any entry in the switch's flow table.

2. The switch then sends a request to the controller to have a new flow table entry created for it.

3. The controller then provides the switch with the data that is to be placed into the flow table's new forwarding entry.

4. The switch uses the new information from the controller to update its flow table.

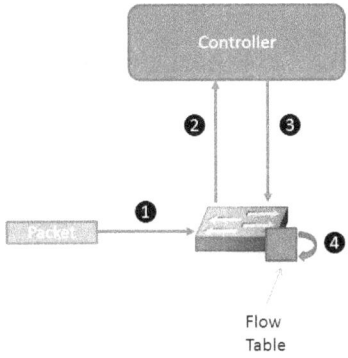

Figure 56: SDN Network Flow Setup Process

Researchers have studied the delays that are imposed on the flow setup process by each step. They report that a controller running on a commodity server should be able to respond to a flow setup message from an SDN switch within a millisecond even when the controller is handling several 100,000 such requests per second [34].

Researchers believe that controller designs should be able to keep up with SDN network growth in order to support reactive flow setup. The design of the control logic will be what ultimately determines the scalability of the controller. One other problem with an SDN network is that as the number of flows in the network increases, the flow table size in each of the switches may be exceeded. Researchers suggest that the ability to install aggregate flow table rules that match a large number of very small flows in order to reduce the load on the controller may address this issue.

Every network has to deal with the challenge of how it reacts to failures in the network and how long it takes it to recover from a failure ("convergence"). SDN networks are no different in this regard. Initial SDN networks used a single controller and this only served to make the resiliency issue an even bigger deal.

If a controller in an SDN network fails, the system is going to have to be able to discover how it wants to recover. The other controllers will need to be able to discover that the controller has failed and as long as there is a state-synchronized backup controller no network state data will be lost. When a controller fails, its connection to the switches that it is managing will be terminated by the switches themselves. This means that the switches will need to be able to support a discover mechanism that will allow them to discover the backup controller as long as it is within the switch's partition.

In an SDN network, how the network reacts to a failure is different from how a legacy network reacts to a network failure. The following five steps, shown in Figure 56, are executed upon failure [34]:

1. The SDN switch will detect that a network failure condition has occurred.

2. The SDN switch will then send a message to the controller notifying it that a network failure has occurred.

3. Once the controller has been notified of the failure, it will use its knowledge of the network to compute new flows that don't use the failed component.

4. The updated flow table information will be pushed out to the affected switches by the controller.

5. The SDN switches that are affected by the failure will receive the updated flow table information from the controller and will then use it to update their flow tables.

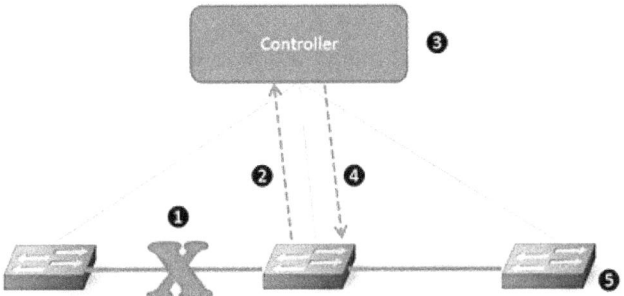

Figure 57: How A SDN Network Converges On a Link Failure

One of the major differences between a SDN network and a legacy network is that in a legacy network when a failure occurs, information about the failure is flooded out to the network. In a SDN network, the information is only sent to the controller. What this means is that the time that it takes to notify the network that a failure has occurred is the same in both legacy and SDN networks.

Researchers believe that in a SDN network, the failure recovery process is very similar to that of a legacy network. What this means is that both types of networks share the same scalability issues. Similar techniques can be used in both types of networks in order to minimize network downtime.

11.5.4 Network Types

Different type of networks pose different challenges for SDN technology. Data centers represent a significant challenge for SDN networks. A single data center that supports virtualization can contain tens of thousands of switching elements. This represents a threat to any centralized controller because the number of control messages that could be generated could easily overwhelm a controller.

In order to prevent this from happening, the solution is to install rules on switches in the data center before they are needed. Doing this will eliminate the situation in which a switch receives a packet that it does not know what to do with it and eliminates the need for the switch to contact the controller. A data center is an ideal network for supporting a distributed controller because of the availability of computing resources. The ample bandwidth means that synchronizing the network views of multiple controllers should be simple to do.

The networks that have been created by service providers look much different from those of data centers. Service provider networks are not nearly as equipment dense as a data center network is and a service provider network will potentially be spread out over a very large distance. Just the size of a service provider network can cause concerns about introducing latency in the updating of a centralized SDN controller. Researchers propose that such a large network could be broken up into separate regions which would each be managed separately by its own controller. The collection of controllers can then synchronize their network views in order to create one common network view.

It is assumed that a service provider's network would contain a large number of flows. In order to permit the SDN switches to manage a large number of flows, researchers recommend that flows be aggregated by the switches in order to minimize the number of table entries that have to be supported.

11.5.5 Next Steps

SDN networks provide network designers with a new set of both opportunities to perform network functions that have never been done before and challenges that will have to be overcome before the true benefits of using SDN technology can be realized. In order for a controller to manage a complete network, a great deal of low level network information has to be provided to it via the OpenFlow interface. The applications that will be built on top of the controller do not need to be exposed to all of this low level information. Instead, they can operate using the abstracted view of the network that the controller will create and present to them. Hiding the details of the network will allow the higher level applications to scale with the network.

SDN technology is new and because of that many new networks are being built for the first time. This means that providing network designers with the ability to test different SDN network configurations will be a critical need. The use of a single centralized controller greatly simplifies the process of testing the network under different configurations in order to determine how it will react. Any SDN testing tools that are developed will need to be able to operate on networks of varying scales. This is an area of research that will be active for a long time.

SDN networks will eventually have to support a wide variety of other networking protocols. The APIs that are being used in SDN networks will have to be modified in order to support additional protocols [34].

The conclusion that researchers have reached is that SDN networks do not pose any more of a scalability issue than legacy networks do. It is believed that the scalability issues that SDN does have can be resolved without taking away any of the novel functionality that SDN offers.

11.6 Network Management Languages For Software Defined Networks

The arrival of SDN networks opens the doors to network monitoring and control functions that have never before been available to network operators. However, the primary protocol that is used in SDN networks is the OpenFlow protocol which allows a centralized controller to control network switches at a very granular level. From a network control point-of-view, this is a powerful feature. However, programming the network becomes much harder because of the large number of operations that need to be executed and the high probability that some important step might be missed. A better way to program SDN networks is needed.

The OpenFlow protocol provides the network programmer with the ability to manipulate the state of the network devices. Creating programs that simultaneously modify how the network performs routing and access control is challenging. This task becomes even more difficult when one takes into account the additional challenges that must be addressed when the network switches are asynchronously updated with changed data. Researchers have labeled the writing of applications for today's SDN networks as being "... *a tedious exercise in low-level distributed programming.*" [36]

The Frenetic research project has been created to design simple abstractions for programming SDN networks. The goal of Frenetic is to provide network programmers with the ability to easily control the three main stages of SDN network management:

1. Monitoring of network traffic
2. Specifying and creating packet forwarding policies
3. Updating network policies in a consistent way

Frenetic accomplishes this by offering a set of declarative abstractions. These can be used by the programmer to query the state of the network, define network forwarding policies, and update the network polices in a consistent manner [36]. A very important characteristic of the Frenetic approach is that it permits individual policies to be created in isolation. Once created, these policies can then be combined with other polices in order to create sophisticated network instructions.

Frenetic views programming the network as being a three part "control loop". The first part of this loop is where the network state is queried. Applications are provided with the ability to subscribe to updates on the state of the network. How the updates are created is left to the runtime system. Applications can be provided with traffic statistics and topology changes which are based on the results of the runtime system collecting switch OpenFlow counters, switch statistics, and event messages.

The next step in the control loop is the reconfiguration of the network based on the state information that has been provided to the application. Ultimately, the network application will specific what packet forwarding behavior a switch should exhibit; however, the way that it does it is via a high-level policy language that is provided by Frenetic.

The final step in the control loop is the reconfiguration of the network. Once again abstractions are used to permit the network application to be able to update individual switches without having to require that the network programmer specify the contents of each OpenFlow flow table entry on each switch. It is the responsibility of the runtime system to ensure network consistency and make sure that the network uses either the old network configuration or the new network configuration but never a combination of the two.

11.6.1 How To Query The Network State Using Frenetic

There are many different events that can change the state of a SDN network. These include link failures, topology changes, changes in the network traffic load, or even the arrival of a specific packet at a specific switch. The SDN controller is able to monitor the state of the network by polling switches and retrieving the packet counter values associated with each flow table entry. Cleary with even a single switch, there could be a great deal of data that must be monitored. The network programmer does not have to worry about this – it will all be taken care of by the runtime system. The network programmer is free to focus on determining what they want to monitor without having to worry about how it will be done.

Frenetic provides network programmers with the tools that they need in order to control what information they have to work with. Frenetic provides four high level operators [36] to accomplish this:

1. Classifying
2. Filtering
3. Transforming
4. Aggregating

As has been previously mentioned in this book, switches in a SDN network have a limited amount of space to store packet forwarding rules. What this means is that it is not possible for the network programmer to configure how a switch should react to every type of packet that it might receive before packet processing starts.

Instead, the Frenetic system allows the network programmer to specify how a switch should react when it receives a packet with a given header. The runtime system will then store this information. When a switch does receive a packet with the header, it will ask the controller what it should do with the packet. The runtime system will instruct the controller to respond with the network programmer's provided data. The switch will then place that information into its limited forwarding rules table and process the packet using it.

Frenetic allows the network programmer to easily collect network statistics. Instead of forcing the programmer to have to be constantly requesting that switches be polled to get updated statistics, Frenetic allows the network programmer to specify an interval for when statistics should be collected. The runtime system will then take care of the task of collecting and presenting the statistics to the network application.

11.6.2 How To Create Network Policies Using Frenetic

Network policies dictate how the network performs tasks such as monitoring, access control, and routing. One of the most significant challenges associated with network polices is when they are created by different individuals or applications and then applied to the same network. Conflicts can occur. Frenetic has developed a network policy language that has a number of features that have been created to allow policies to be created and combined in a modular fashion.

When a SDN switch is powered up, every time it receives a packet it won't know where to forward the packet to. This means that each packet will then be sent to the controller. The controller will then do one of two different things. The first is to decide that the controller does not need to see any of the other packets that will be in that flow. In this case, the controller will provide the switch with the forwarding rule that will apply actions to all future packets that belong to this flow.

However, in the case that the controller does want to process other packets in the flow, it will instruct the switch what to do with the current packet. However, no rules will be installed in the switch's flow table. This means that the next time that a packet with the same header is received by the switch, it will once again be forwarded to the controller.

This selective processing of flow packets allows high-level policy rules to be expressed in low-level switch-level actions.

11.6.3 How To Perform Consistent Updates Using Frenetic

Modern networks experience a great deal of change over time. Links fail, network traffic increases, maintenance occurs. When these changes happen, the network policy that is currently being used by the network needs to be changed. This opens up the possibility that changing the network policy could result in forwarding loops, security breaches, or transient outages. In order to avoid these conditions, changes must be applied to the network in a consistent manner.

Frenetic supports what researchers call a "per-packet consistent update" [36]. What this means is that if the network starts to process a packet using network policy set #1 and while the packet is traversing the network the network policy is updated to be policy #2, then the packet will be processed using policy #1 throughout its journey. This ensures that no loops or other conflicts will occur. Since the Frenetic system takes care of this, the network programmer can assume that they can create applications that are both reliable and dynamic.

The Frenetic system ensures that per-packet consistency is achieved using a two-phase update system that marks packets [36]. When a packet is received into the network, it is stamped with the version of the network policy that is in effect when the packet was received. As the packet travels through the rest of the network, this version number is tested at each switch that processes the packet. Each switch in the network has two copies of the network policy – the current one and the previous one. The switch has the ability to process a packet using either policy. The decision regarding which network policy to use is made based on the packet's policy stamp. Once all of the packets that were stamped with the old policy information have left the network, the controller visits each switch, deletes the old policy information, and updates the switch to use the new policy information.

The Frenetic system also supports a per-flow consistent update [36]. This ensures that a stream of packets that make up a flow are all processed using the same network policy. In order to implement this feature, the runtime system can use soft timeouts to cause the rules that were processing the old configuration to be removed and to be replaced with pre-installed rules for the new configuration.

The ability to program a SDN network is going to be a critical feature of future networks. In order to realize the true potential of SDN it is going to be necessary to be able to easily create network applications that can make use of SDNs network topology information and consistently update the switches in the network. The Frenetic project has created a way to accomplish this. The challenges of abstracting how the SDN can be queried and updates made in a consistent way in order to avoid conflicts have been solved. There is much research still to be done, but the foundation of tomorrow's SDN network applications has already been laid.

11.7 Creating A SDN Controller That Is Both Elastic And Distributed

One of the great advantages of a SDN network is that it can provide a single, unified view of the entire network. This view is provided by and maintained by a centralized controller.

However, the use of a centralized controller quickly brings up the two issues of scalability and reliability – can the controller keep up with the growth of the network? Researchers have been looking into how best to create a logically centralized, but physically distributed controller that can solve both of these problems.

One of the key issues with any distributed controller is that it may be possible for any one component of the controller to become overwhelmed by network traffic. In many distributed controller designs, network SDN switches are assigned to a given controller. This static mapping creates two issues. If a given controller becomes overloaded due to a network traffic surge, then it may not be able to process all of the flows that it is responsible for. Additionally, the reverse situation may arise. The controller may become underutilized and be able to support more switches. Neither one of these scenarios is ideal.

In order to address these distributed controller issues, researchers have developed a distributed controller architecture called ElastiCon [37]. In this architecture, the load on a controller is dynamic: an overloaded controller can have its load shifted to another controller. Additionally, more controllers can be brought online in the event that additional controller capacity is needed. If the load on the controller shrinks, then controllers can be removed from the system.

In order to create a distributed controller that can adapt to the load that is currently being applied to it, there are three operations that will have to be supported by the system:

1. The system will need to periodically be load balanced in order to achieve the optimal switch to controller ratio.

2. When the load grows larger than can be handled by the current pool of controllers, additional controllers will have to be added to the system. Additionally, once the new controller has been added to the system, switches will have to be migrated to the new controller.

3. When the load falls below a predefined level, then the pool of controllers will need to be shrunk. Those switches that are being managed by the controllers that are being removed will have to be migrated to controllers that are remaining active.

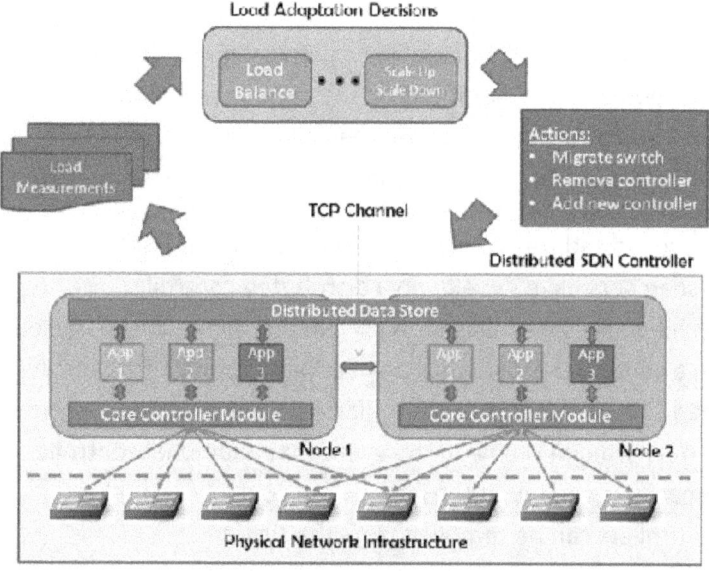

Figure 58: Architecture Of The ElastiCon Distributed Controller

The major components that make up the Elasticon distributed controller are shown in Figure 58 [37]. The architecture consists of the following components:

- Multiple **controller nodes** that communicate with each other in order to control the network

- Each controller node has a **core control module** which is responsible for executing all of the functions that a centralized controller would execute. These functions include connecting to a SDN switch and management of the communications that occur between a SDN switch and a higher level application. This module is also responsible for working with other controllers in order to determine which controller is to be the master controller. It will also handle the logic that is required in order to move switches between controllers.

- **Physical Network Infrastructure** which consists of the SDN switches and the links that are used to carry network traffic. Each switch can connect to one or more nodes. In the case that a switch is connected to multiple controllers, one controller will be the master and the rest will be slaves.

- The **distributed data store** is responsible for creating a logical single centralized controller. This database stores all of the SDN switch specific information that will be needed by the distributed controllers.

- The **TCP Channel** is used by each controller to talk with other distributed controllers in order to communicate with a switch that is attached to another controller or to coordinate the transfer of a switch to another controller.

- The **application modules** are responsible for implementing the controller's network application logic. These applications are responsible for managing the switches that have been assigned to the controller. The current state of each application will be kept in the distributed data store to help with the switch migration process and as an aid in the event of a distributed controller failure.

11.7.1 How To Migrate Switches Using The ElastiCon Distributed Controller

Load balancing will need to be performed periodically in order to ensure that none of the distributed controllers become overloaded. The OpenFlow protocol currently does not have any mechanism built-in to perform this function.

As has been previously discussed in this book, the OpenFlow protocol supports three different roles for distributed controllers that are connected to a switch. The roles are as follows:

1. EQUAL
2. SLAVE
3. MASTER

The protocol specifies that only one controller can be in the Master role at any time. Controllers in the Master and the Equal roles are permitted to modify the state of the switch and will receive asynchronous messages from the switch. A controller in the Slave role has read only access to the switch and will not receive asynchronous messages from the switch.

In studying how to implement a switching protocol, researchers identified two properties that had to be guaranteed at all times during the switching process:

1. There must always be a controller that is connected to the SDN switch that is currently in either the Master or the Equal mode. The controller that is in the Master or Equal mode must always be allowed to complete the processing of an asynchronous message before the controller's role is changed.

2. Only one controller will be permitted to process asynchronous messages from the SDN switch at any time.

The challenge of implementing the switch migration protocol comes from the scenario in which the SDN switch has sent an asynchronous message to a controller that it is attached to which is currently in the Master role. If that controller has not yet replied to the switch, then the switch cannot be migrated to another controller because it would violate the first rule. Simply changing the role of the Master controller to be Equal and then making the switch's new controller be in the Master role could potentially violate rule 2 [37].

The researchers who were investigating how to extend the OpenFlow protocol in order to support the migration of switches eventually decided to implement a trigger event in order to facilitate the communication of the migration between two controllers. The process of transferring a switch from one controller to another without violating either of the rules requires the following four steps [37]:

1. **The target controller has its role changed to be Equal**: The first step of the migration process starts with the controller who currently has the Master role for the switch sending a command to the controller who will be receiving the switch via the TCP channel to change its mode to Equal.

   The controller that will be receiving the switch then sends a command to the switch telling it to change its role to be Equal. Once this is completed, the switch informs the receiving controller that it is now in the Equal role. The receiving controller then informs the Master controller that the role change has been completed.

2. **Create A Dummy Flow**: In order to coordinate the migration of the SDN switch from the sending controller to the receiving controller, the sending controller will now command the switch to establish a dummy flow. Both controllers will have agreed before what the dummy flow will look like. The dummy flow will be constructed so that it will never match an incoming packet header. The switch will acknowledge to the sending controller that the dummy flow has been established. Note that since the receiving controller is in the Equal mode, it receives copies of all of the messages that the switch sends to the sending controller.

   At this point in time the sending controller will send the switch what is called a "barrier message". The purpose of this message is to make sure that there is no possible way that the switch could possibly process the delete message for the dummy flow before it processes the insert message for the dummy flow. Barrier request/reply messages are used by the controller to ensure message dependencies have been met or to receive notifications for completed operations.

Once the sending controller has this acknowledgement, the sending controller will now command the switch to delete the dummy flow. The switch will delete the dummy flow and will then send a confirmation message to the sending controller. The receiving controller will also receive a copy of this confirmation message from the switch. At this point in time the receiving controller is now in control of the switch and the sending controller will no longer send any configuration commands to that switch.

3. **Use A Barrier Message To Flush Pending SDN Switch Requests**: Once the migration of the switch from the sending controller to the receiving controller has occurred, the migration process is not yet considered to be complete. Specifically, the sending controller may still have some packets that the switch is expecting it to provide instructions on how to handle.

In order to ensure that this situation has been cleared up, the sending controller will transmit a barrier message that, according to the OpenFlow protocol will cause the following events to occur: first, messages that the switch has received before the barrier message must be fully processed before the barrier message, including sending any resulting replies or errors. Next, the barrier message must then be processed and a barrier reply sent. Finally, messages after the barrier may then begin processing.

After the sending controller receives the barrier reply message from the switch, the sending controller will send an end of migration message to the receiving controller via the TCP channel.

4. **Change Receiving Controller To Master Mode**: In the final stage of the switch migration process, the receiving controller tells the SDN switch to set its mode to be Master. When the switch does this, it will automatically set the mode of the sending controller to be Slave. At this point all future autonomous messages from the switch will only be seen and processed by the receiving controller. The receiving controller will update the distributed data store to indicate that it is now the master of the switch that has been migrated.

11.7.2 Load Adaptation

An important part of the ElastiCon Distributed Controller model is its ability to perform load adaptation. As shown in Figure 57, there are three parts to load adaptation [37]:

1. Load measurement

2. Adaptation decision computation

3. Migration action

Distributed controller load statistics are continuously calculated by a load estimation module that executes as part of the controller. Load statistics consist of three controller measurements: CPU usage, memory usage, and the rate of network I/O. Researchers have found that the CPU rate is the best measurement of the message arrival rate [37].

Two sets of performance thresholds are established for the ElastiCon Distributed Controller model. The first is a high / low loading limit for individual controllers and the second is a high / low limit for the set of active controllers as a whole. When a given controller exceeds its high threshold but the overall load on the set of active controllers has not exceeded its threshold, then load balancing will occur. Load balancing will result in switches that are being managed by the overloaded controller being migrated to other controllers who are operating under their high thresholds.

If the high or low limits for the set of active controllers has been exceeded, then either additional controllers will be activated or active controllers will be deactivated. Both the load conditions and the topology of the network will be taken into account when determining which controllers will be set to master mode.

When a new controller is added to the set of distributed controllers, the switches that are being managed will not know how to communicate with it. The OpenFlow protocol allows a controller to set the IP address of other controllers within a switch. A controller which will be migrating the switch to the newly activated controller will perform the IP address update and then the switch migration procedure will be performed.

12 Conclusion

The opportunity to participate in a revolution in the field of Information Technology is a rare treat. The birth of mainframes, the arrival of client-server computing, the dawn of the Internet era, these are all remembered today as being pivotal points in the evolution of computing. The arrival of Software Defined Networking (SDN) appears to once again be providing us with a chance to be part of something new and wonderful.

Although everyone loves a new technology just because it is shiny and new, SDN has a lot more going for it than just being new. The networks that we are building using today's technology have gotten out of hand. They are just too big and too complicated for us to manage in any way that will allow us to keep our costs under control. Everyone knows that a better way is needed. The question that is being asked right now is if SDN is that better way?

More research will be required, but what we know so far looks very promising. SDN is simple: its fundamental philosophy in which the network's control plane is separated from its data plane is very easy to grasp. The advantages that SDN appears to offer network operators include better utilization of expensive network links, service engineered paths, and service appliance pooling, etc. are all things that everyone can agree that they want.

All too often new networking ideas look great on paper but for reasons that only become clear later on they don't work out when they are applied to the real world. That's one of the things that makes SDN so exciting. It looked so good on paper that Google was willing to take the lead and transform one of the two mission-critical backbone networks that support their business so that it now uses SDN technology.

What makes this level of commitment by Google even more amazing is that the OpenFlow protocol is still being refined and when Google moved forward with the project to transform their network no commercially available OpenFlow servers were available for them to use – they had to build their own. Having made this investment in both time and engineering talent, Google has told us that they have now been able to achieve things with their new SDN network that were not possible before.

If Google can do it the hard way, why can't the rest of us do the very same thing? It has become easy to get the three things that we'll need to build our own SDN networks: the OpenFlow protocol is now more mature, we can purchase simple routers based on merchant silicon that support the OpenFlow protocol, and controllers are now available both from multiple firms and as open-source projects.

Today we are investigating what SDN can do for our packet switched networks. This is only the start. We have so many other types of networks from optical transport to wireless that we think can also benefit from what SDN has to offer us. The true power of SDN will only be realized when we can take our multilayered networks, fully virtualize each layer, and have them work closely together in order to deliver a level of service to our applications that we've never before been able to dream of.

What makes SDN hold so much promise for the future is not what it provides for us today which is a great deal, but rather what it can make possible for our tomorrow. For the first time

SDN allows us to virtualize the entire network. We will finally have a complete model of the network with which to play. This means that those really smart programmers out there will now have something else to tinker with.

None of us can image what the future holds. With the ability to present data that shows a complete model of what the network is both capable of and what it is currently doing to novel new software applications will open many new doors. Those applications will be able to use this data to create insights to that data that we have previously only dreamed of having. What's even more important is that through the tools that SDN technology provides us with, including the OpenFlow protocol, those applications will now be able to program the network to do what they need it to do. Just think of the possibilities!

The future of SDN has not been written yet. Will it remain an open standard in which freely available software will be used with simple high-speed switches to create networks that are unlike anything that we have today? Or will today's dominate vendors create unique solutions that customers feel more comfortable in adopting because they know that it will work with the equipment that they already have and they know where to go to get support if something doesn't work for them? Only time will tell.

The one thing that we do know is that we are only at the beginning of the SDN story. The networks that we will be building tomorrow will look nothing like the networks that we have already built. This means that we are going to need to dream up a new set of operational procedures, network management techniques, and security features. All of the solutions that we've created in the past were for yesterday's networks. Tomorrow's networks will require us to solve the same problems all over again. Looks like it's time to get busy – we've got a lot of work to do!

13 Reference

[1] The History of the Mainframe Computer
http://␣ww.vikingwaters.com/htmlpages/MFHistory.htm

[2] Design and Implementation of a Routing Control Platform
http://www.cs.princeton.edu/~jrex/papers/rcp-nsdi.pdf

[3] SANE/inSANE: Designing Secure Networks from the Ground-Up
http://yuba.stanford.edu/sane/

[4] Power, Pollution and the Internet
http://www.nytimes.com/2012/09/23/technology/data-centers-waste-vast-amounts-of-energy-belying-industry-image.html

[5] ONS 2012 Keynote : Urs Hoezle, SVP Google
http://www.youtube.com/watch?v=JMkvCBOMhno

[6] Software Defined Networks: A Carrier Perspective - Stuart Elby, Verizon
http://www.youtube.com/watch?v=xsoUexvljGk

[7] SDN for Service Provider Networks: Technology, Applications, and Markets
Programmable WAN is SFW, David Ward

http://www.youtube.com/watch?v=GPZ9ZcruHo8

[8] How Software Defined Networks/OpenFlow Can Transform Network Performance, Charles Ferland
http://www.youtube.com/watch?v=VjV5_23TUzw

[9] Insieme FAQ: a few key facts
http://www.networkworld.com/news/2013/110613-insieme-faq-275585.html?page=2

[10] Is Cisco's SDN Architecture Really That Special?
http://www.sdncentral.com/news/is-cisco-sdn-architecture-really-that-special/2013/11/

[11] ONS2012 Amin Vahdat - SDN Stack for Service Provider Networks
http://www.youtube.com/watch?v=ED51Ts4o3os

[12] MKM: Cisco Biggest Loser in AT&T SDN Plans
http://www.lightreading.com/carrier-sdn/sdn-architectures/mkm-cisco-biggest-loser-in-atandt-sdn-plans/d/d-id/707122?goback=.gde_4359316_member_5823678536455581696#!

[13] "Google shares lessons learned as early software-defined network adopter",
Network World, April 18, 2012, Colin Neagle
http://www.networkworld.com/news/2012/041812-google-

openflow-258406.html?hpg1=bn

[14] "Google's software-defined/OpenFlow backbone drives WAN links to 100% utilization"
Network World, June 07, 2012, John Dix
http://www.networkworld.com/news/2012/060712-google-openflow-vahdat-259965.html

[15] The [Xen] Hypervisor
http://www.xenproject.org/developers/teams/hypervisor.html

[16] EPA Report on Server and Data Center Energy Efficiency
http://www.energystar.gov/index.cfm?c=prod_development.server_efficiency_study

[17] How to Find Virtualization Software & Hypervisors
http://virtualization.findthebest.com/guide

[18] SNIA TECHNICAL TUTORIAL: Storage Virtualization
https://www.snia.org/education/storage_networking_primer/stor_virt

[19] White Paper: "Software-Defined Networking: The New Norm for Networks"
https://www.opennetworking.org/images/stories/downloads/sdn-resources/white-papers/wp-sdn-newnorm.pdf

[20] OpenFlow Switch Specification, Version 1.4.0 (Wire Protocol 0x05), October 14, 2013

https://www.opennetworking.org/images/stories/downloads/sdn-resources/onf-specifications/openflow/openflow-spec-v1.4.0.pdf

[21] ONS 203 Keynote address by Vint Cert
https://www.youtube.com/watch?v=ZrUGythq9TI

[22] OpenFlow Management and Configuration Protocol (O F- C o n fi g 1.1.1)
https://www.opennetworking.org/images/stories/downloads/sdn-resources/onf-specifications/openflow-config/of-config-1-1-1.pdf

[23] OpenFlow™ Conformance Testing Program
https://www.opennetworking.org/sdn-resources/onf-specifications/openflow-conformance

[24] SDN showdown: Examining the differences between VMware's NSX and Cisco's ACI
http://www.networkworld.com/news/2014/010614-vmware-nsx-cisco-aci-277154.html

[25] Virtual network overlay scalability: Legit problem or trolling?
http://searchsdn.techtarget.com/news/2240205116/Ciscos-response-to-VMware-NSX-and-the-future-of-SDN

[26] Juniper open sources Contrail SDN software stack
http://www.theregister.co.uk/2013/09/16/juniper_contrail_sdn_controller_ships/

[27] Juniper Launches Contrail SDN Software, Goes Open Source
http://www.networkcomputing.com/data-networking-management/juniper-launches-contrail-sdn-software-g/240161354

[28] The OpenDaylight Project
http://www.opendaylight.org

[29] Big Switch uncloaks, fires virty network wares at VMware/Nicira
http://www.theregister.co.uk/2012/11/13/big_switch_networks_sdn/

[30] Incremental SDN Deployment in Enterprise Networks
Dan Levin, Marco Canini, Stefan Schmid, Anja Feldmann
Proceedings of the SIGCOMM 2013, pp.473-474

[31] Software Define Networking Opportunities For Transport
Dave McDysan
IEEE Communications Magazine, March 2013, Vol. 51, No. 3, pp. 28-31

[32] Extending Software Defined Network Principles to Include Optical Transport
Steven Gringeri, Nabil Bitar, and Tiejun J. Xia
IEEE Communications Magazine, March 2013, Vol. 51, No. 3, pp. 32-40

[33] Achieving High Utilization with Software-Driven WAN
Chi-Yao Hong, Srikanth Kandula, Ratul Mahajan, Ming Zhang,

Vijay Gill, Mohan Nanduri, Roger Watterhofer
Proceedings of the SIGCOMM 2013, pp.15-26

[34] On Scalability of Software-Defined Networking
Soheil Yeganeh, Amin TooToonchian, Yashar Ganjali
IEEE Communications Magazine, February 2013, Vol. 51, No. 2, pp.136-141

[35] A. Tootoonchian et al., "On Controller Performance in Software-Defined Networks," Proc. USENIX Hot-ICE '12. 2012. pp. 10-10

[36] Languages for Software Defined Networks
Nate Foster et al.
IEEE Communications Magazine, February 2013, Vol. 51, No. 2, pp.128-134

[37] Towards an Elastic Distributed SDN Controller
Advait Dixit et al.
Proceedings of the SIGCOMM 2013, pp.573-578

Photo Credits:

Cover - Rosmarie Voegtli
https://www.flickr.com/photos/rvoegtli/

Other Books By The Author

Product Management

- What Product Managers Need To Know About World-Class Product Development: How Product Managers Can Create Successful Products

- How Product Managers Can Learn To Understand Their Customers: Techniques For Product Managers To Better Understand What Their Customers Really Want

- Product Management Secrets: Techniques For Product Managers To Boost Product Sales And Increase Customer Satisfaction

- Product Development Lessons For Product Managers: How Product Managers Can Create Successful Products

- Customer Lessons For Product Managers: Techniques For Product Managers To Better Understand What Their Customers Really Want

- Product Failure Lessons For Product Managers: Examples Of Products That Have Failed For Product

Managers To Learn From

- Communication Skills For Product Managers: The Communication Skills That Product Managers Need To Know How To Use In Order To Have A Successful Product

- How To Have A Successful Product Manager Career: The Things That You Need To Be Doing TODAY In Order To Have A Successful Product Manager Career

- Product Manager Product Success: How to keep your product on track and make it become a success

Public Speaking

- Tools Speakers Need In Order To Give The Perfect Speech: What tools to use to create your next speech so that your message will be remembered forever!

- How To Create A Speech That Will Be Remembered

- Secrets To Organizing A Speech For Maximum Impact: How to put together a speech that will capture and hold your audience's attention

- How To Become A Better Speaker By Changing How You Speak: Change techniques that will transform a speech into a memorable event

- How To Give A Great Presentation: Presentation techniques that will transform a speech into a memorable event

- How To Rehearse In Order To Give The Perfect Speech: How to effectively rehearse your next speech to that your message be remembered forever!

- Secrets To Creating The Perfect Speech: How to create a speech that will make your message be remembered forever!

- Secrets To Organizing The Perfect Speech: How to organize the best speech of your life!

- Secrets To Planning The Perfect Speech: How to plan to give the best speech of your life

- How To Show What You Mean During A Presentation: How to use visual techniques to transform a speech into a memorable event

CIO Skills

- Becoming A Powerful And Effective Leader: Tips And Techniques That IT Managers Can Use In Order To Develop Leadership Skills

- CIO Secrets For Growing Innovation: Tips And Techniques For CIOs To Use In Order To Make Innovation Happen In Their IT Department

- Your Success As A CIO Depends On How Well You Communicate: Tips And Techniques For CIOs To Use In Order To Become Better Communicators

- What CIOs Need To Know About Working With Partners: Techniques For CIOs To Use In Order To Be Able To Successfully Work With Partners

- Critical CIO Management Skills: Decision Making Skills That Every CIO Needs To Have In Order To Be Able To Make The Right Choices

- How CIOs Can Make Innovation Happen: Tips And Techniques For CIOs To Use In Order To Make Innovation Happen In Their IT Department

- CIO Communication Skills Secrets: Tips And Techniques For CIOs To Use In Order To Become Better Communicators

- Managing Your CIO Career: Steps That CIOs Have To Take In Order To Have A Long And Successful Career

- CIO Business Skills: How CIOs can work effectively with the rest of the company!

IT Manager Skills

- Save Yourself, Save Your Job – How To Manage Your IT Career: Secrets That IT Managers Can Use In Order To Have A Successful Career

- Growing Your CIO Career: How CIOs Can Work With The Entire Company In Order To Be Successful

- How IT Managers Can Make Innovation Happen: Tips And Techniques For IT Managers To Use In Order To Make Innovation Happen In Their Teams

- Staffing Skills IT Managers Must Have: Tips And Techniques That IT Managers Can Use In Order To Correctly Staff Their Teams

- Secrets Of Effective Leadership For IT Managers: Tips And Techniques That IT Managers Can Use In Order To Develop Leadership Skills

- IT Manager Career Secrets: Tips And Techniques That IT Managers Can Use In Order To Have A Successful Career

- IT Manager Budgeting Skills: How IT Managers Can Request, Manage, Use, And Track Their Funding

- Secrets Of Managing Budgets: What IT Managers Need To Know In Order To Understand How Their Company Uses Money

Negotiating

- Learn How To Signal In Your Next Negotiation: How To Develop The Skill Of Effective Signaling In A Negotiation In Order To Get The Best Possible Outcome

- Learn The Skill Of Exploring In A Negotiation: How To Develop The Skill Of Exploring What Is Possible In A Negotiation In Order To Reach The Best Possible Deal

- Learn How To Argue In Your Next Negotiation: How To Develop The Skill Of Effective Arguing In A Negotiation In Order To Get The Best Possible Outcome|

- How To Open Your Next Negotiation: How To Start A Negotiation In Order To Get The Best Possible Outcome

- Preparing For Your Next Negotiation: What You Need To Do BEFORE A Negotiation Starts In Order To Get The Best Possible Deal

- Learn How To Package Trades In Your Next Negotiation

- All Good Things Come To An End: How To Close A Negotiation - How To Develop The Skill Of Closing In Order To Get The Best Possible Outcome From A Negotiation

Miscellaneous

- The Internet-Enabled Successful School District Superintendent: How To Use The Internet To Boost Parental Involvement In Your Schools

- Power Distribution Unit (PDU) Secrets: What Everyone Who Works In A Data Center Needs To Know!

- Making The Jump: How To Land Your Dream Job When You Get Out Of College!

- How To Use The Internet To Create Successful Students And Involved Parents
-

The Revolution In Network Design And How It Affects You

> This book has been written with one goal in mind – to show you how the revolution in network design called Software Defined Networking (SDN) is going to affect you! This changes everything, let's makes sure that you are ready.
>
> **Let's Prepare Your Network For The Future!**

What You'll Find Inside:

- **THE IMPORTANCE OF SDN**

- **HOW TELECOM SERVICE PROVIDERS VIEW SDN**

- **THE DEVELOPER AND THE NETWORK**

- **EXAMPLES OF NETWORKED APPLICATIONS THAT CAN ONLY BE OFFERED IN AN SDN NETWORK**

- **GOOGLE AND SDN**

Dr. Jim Anderson brings his 25 years of real-world experience to this book. He's spent over 25 years working in the Telecommunications industry and teaching at Universities. His insights will show you how SDN will impact your life and your network.

www.ingramcontent.com/pod-product-compliance
Lightning Source LLC
Chambersburg PA
CBHW070229190526
45169CB00001B/132